LEAPFROGGING DEVELOPMENT IN EMERGING ASIA: CAUGHT BETWEEN GREENING AND POLLUTION

Leapfrogging Development in Emerging Asia: Caught between Greening and Pollution

Peter Ho
Editor

Nova Science Publishers, Inc.
New York

For permission to use material from this book please contact us:
Telephone 631-231-7269; Fax 631-231-8175
Web Site: http://www.novapublishers.com

LIBRARY OF CONGRESS CATALOGING-IN-PUBLICATION DATA

Ho, Peter.
Leapfrogging development in emerging Asia : caught between greening and pollution / Peter Ho.
p. cm.
ISBN 978-1-60456-410-5 (hardcover)
1. Environmental protection--Asia. 2. Pollution--Asia. I. Title.
TD171.5.A7842H6 2008
363.70095--dc22

2008000853

Published by Nova Science Publishers, Inc. New York

CONTENTS

PREFACE

Many foreign observers foresee that Asia will dominate the world economy for many decades to come. However, the development of Taiwan, Singapore, Hong Kong and Korea, and, recently, the emerging economies of China and Vietnam, has also come at a price – environmental degradation. This edited volume will focus on the relation between environment and development in an Asian context. More specifically, the book seeks to answer the question whether Asia's newly emerging economies can find cleaner development trajectories than Western economies by "leapfrogging" the dirty stages of development. In so doing, this book will be the first work that extends the debate on environmental leapfrogging to the Asian continent.

Asia is facing great environmental challenges from urbanization and industrial growth. To avoid an imminent environmental crisis, the emerging economies in Asia need to improve their energy efficiency, develop nonfossil energy resources, clamp down on pollution, and encourage technologies that minimize pollution. A critical question in this respect is: how can the state ensure the *greening* of industries and business without inhibiting economic growth?

Some scholars argue that the emerging Asian countries might in fact be situated at a unique juncture: they have an unparalleled opportunity to find more environmentally friendly development paths by learning from the earlier, industrialized nations. With the right policy mix, emerging economies might achieve rapid economic development while avoiding environmental degradation on the scale of that created by the United States, European Union and other older industrialized nations. "Doing it right the first time" – by installing clean, efficient technologies, developing the institutional capacity, and the appropriate governance structure – could lead to "leapfrogging" development, and can build economies that are competitive and more sustainable than those with an older industrial base.[1]

In the academic literature on greening, the implementation of decentralized, integrated, voluntary and market-based approaches is deemed critical for the environmental governance of the industrial sector. The appeal of environmental leapfrogging induced several Asian economies to adopt such measures, at times under the auspices of "good governance" programs of multilateral and bilateral donors. The book brings together several research articles that present empirical studies, as well as theoretically inspired work around the critical issues above.

---

[1] Luc Soete, (1985), "International Diffusion of Technology, Industrial Development and Technological Leapfrogging," *World Development*, Vol. 13, No. 3, pp. 409-22.

This book is written by a team of leading experts in their field, ranging from development studies, sociology, political studies, and economics. The volume fits in with the larger, international body of studies on environmental management and the greening of industries. This book will be of interest to a wide readership of students and professionals concerned with development in contemporary Asia, with particular reference to environmental studies, industrial pollution control, social movements, and developmental state theory.

The Editor

Peter Ho is professor of International Development Studies at the Faculty of Spatial Sciences and director of the Centre for Development Studies of the University of Groningen (www.rug.nl/cds). He has published widely on issues of environmental policy, collective action and property rights, and institutional change. His recent books include *Rural Development in Transitional China* (Frank Cass, 2003), *Developmental Dilemmas* (Routledge, 2005), *China's Limits to Growth* (Blackwell, 2006), *Institutions in Transition* (Oxford University Press, 2005) and *China's Embedded Activism* (Routledge, 2008).

In: Leapfrogging Development in Emerging Asia
Editor: Peter Ho, pp. 1-17 ISBN: 978-1-60456-410-5

Chapter 1

GREEN INDUSTRIES, CLEAN ENVIRONMENT? CHINA, VIETNAM, TAIWAN AND SINGAPORE COMPARED

Peter Ho
Centre for Development Studies and Faculty of Spatial Sciences
University of Groningen, The Netherlands

LIST OF ACRONYMS

CFC	Chlorofluorocarbons
DPP	Democratic Progressive Party
FDI	Foreign Direct Investment
GDP	Gross Domestic Product
IMF	International Monetary Fund
IT	Information Technology
KMT	Kuomintang
NIC	Newly Industrializing Country
NIE	Newly Industrializing Economy
SME	Small and Medium-scale Enterprise
TNC	Transnational Company
WRI	World Rcsources Institute

1. STATE, INDUSTRY AND ENVIRONMENT IN NICS

Over the past decades, the world has witnessed the profound transformation of Singapore, Hong Kong, South Korea, and Taiwan from impoverished developing regions into strong and internationally competitive economies. Since 1965, these four economies have quadrupled their share of world production and trade and quintupled their per capita incomes. Between 1965-1986, the per capita GDP in each one grew at least 6 percent per year – by comparison, during the same period, Japan and the United States registered, respectively, annual increases of 4.3 and 1.6 percent. The experience of these four economies, popularly dubbed the four "Little Tigers" of Asia, sparked great academic interest in the study of so-called Newly Industrializing Economies or Newly Industrializing Countries (NIE or NIC). Different scholars tend to label different economies as NICs, or when taking the timing of industrialization into account "first generation" and "second generation" NICs. The general consensus is, however, that various economies in South and Latin America, as well as in Asia could be termed NICs, including Mexico, Brazil, Chile, Argentina, India, Indonesia, Malaysia, and Thailand. In this volume, we will focus on four of these: China, Vietnam, Singapore and Taiwan. It should be noted that for the sake of argument, the NICs here are conceived primarily as economies, rather than countries or nations. For instance, Hong Kong used to be a colony of the United Kingdom until 1997 when it became a special administrative region under China. In addition, Singapore – and formerly Hong Kong – as city-states are not confronted with many of the sharp rural-urban divisions characteristic of most developing countries.

The rapid economic development in the NICs has come at a price. Contrary to their economic successes, the NICs have been much less successful in terms of ecological sustainability and environmental protection. For example, Brazil, Malaysia and Indonesia have abundant natural resources, but those riches – notably their forest and soil reserves – are being rapidly depleted. In addition, mega-cities in Mexico, Argentina and India suffer from increased air pollution, traffic congestion and a boom in the generation of solid waste. It is evident that NICs face new environmental challenges from urbanization and industrial growth. They need to improve their energy efficiency, develop nonfossil energy resources where possible, clamp down on air and water pollution, and encourage new technologies that minimize or prevent pollution. The rapid industrialization in the NICs poses a dual challenge to the state. If economic development is to be sustainable in the long run, the state needs to manage and channel processes of industrialization and the environmental pressures these entail. Simultaneously, the state in NICs is faced with solving major development problems, ranging from poverty alleviation, food security and the protection of natural resources that form the primary base of the economy and a vital source of employment. A critical question in this respect is: how can the state effect environmental reform in the industrial sector without inhibiting economic growth? In other words, in what way can the state in NICs ensure the greening of industries and business while simultaneously safeguarding their competitiveness at the national and international level?

In seeking the answer to this question, some scholars would argue that NICs are in fact situated at a unique juncture: they have an unparalleled opportunity to find different development paths and in so doing to provide models that other countries could follow. Put simply, with the right policy mix, they might achieve rapid economic development while avoiding environmental degradation on the scale of that created by the United States, European Union, Japan, and other earlier industrialized nations. "Doing it right the first time" – by installing clean, efficient technologies as well as developing the institutional capacity and the appropriate governance style to enforce environmental regulations – could lead to "leapfrogging" the development process and building industrial economies that are both competitive and more sustainable than those of economies with an older industrial base. For an effective, more cost-efficient environmental governance of the industrial and business sector, it is often proposed to implement decentralized, integrated, voluntary and market-based approaches. The appeal of leapfrogging induced several NICs to adopt such measures, at times under the auspices of "good governance" programmes of multilateral and bilateral donors.

Against this backdrop, the current edited volume has brought together seven contributions with case studies from first and second generation NICs: China, Vietnam, Taiwan and Singapore.[1] As we will see below, in spite of certain parallels, these four NICs differ from each other in critical ways. As such they might provide important insights into the effectiveness and suitability of various environmental policy measures in the greening of industries. Before turning to a discussion of the chapters, it is necessary to introduce some of the main theoretical concepts and ideas on the greening of industries and business.

2. Green Industries and Environmental Governance

Concepts, Controversies and Discussions

In a now classical work, Albert Weale posited that the failure of industrial pollution control policies in the 1960s and 1970s was due to inherent administrative weaknesses in the United States and Europe. In his words, these policies,

> relied primarily upon the techniques of administrative regulation. They involved selective attention, stressing some problems at the expense of others. (...) They favoured the regulation of pollution by legislation that was specific to receiving medium. They rested on the assumption that environmental policy could be treated as a discrete policy area in its own right. And they allowed problem displacement across time and space.
>
> (Albert Weale, The New Politics of Pollution, 1992)

According to Weale, improved pollution control and effective greening of industries need a fundamental restructuring of the environmental governance style. Whereas pollution control in the '60s and '70s relied on an interventionist state that used adversarial "command-and-control" measures and "end-of-pipe" technologies, the "new politics of pollution" would greatly benefit from the implementation of decentralized, market-based and integrated policy measures. These typically include voluntary schemes of coregulation (e.g. ISO 14000 certification and corporate environmental accounting by firms); the involvement of nongovernmental groups in tri-partite negotiations with state and industries; financial incentive mechanisms such as tradable waste permits, green taxes and subsidies; and cleaner production technologies and concepts of industrial ecology that target the entire chain of production and consumption, rather than just the end of the polluting pipe.[2]

---

[1] The chapters are based on a revised version of papers which appeared as a special issue of the *International Journal of Environment and Sustainable Development*, Vol. 4, No. 3, 2005. The chapter by Khoa and Ho is a new contribution which has been added to this volume.

[2] At a more general level – that of environmental reform in society, economy and polity – Weale's argument has become known as "ecological modernization" as the German thinker Joseph Huber coined it. Huber J., (1982), Die Verlorene Unschuld der Ökologie: Neue Technologien und superindustrielle Entwicklung, Fisher: Frankfurt am Main; and Huber J. (1985), Die Regenbogengesellschaft: Ökologie und Sozialpolitik. Fisher: Frankfurt am Main. See also Mol, Arthur P.J. and D.A. Sonnenfeld (eds.), (2000), *Ecological Modernisation around the World: Perspectives and Critical Debates*, London: Frank Cass.

In certain ways, the strong, interventionist state relying on command-and-control measures as described by Weale resonates with the concept of a "developmental state." In seeking theoretical explanation for the economic wonders of the "Little Tigers" of Southeast and East Asia social scientists have wielded the notions of developmental state theory. In economies of the NICs, the central state has played a decisive role in stimulating and guiding economic development through the pursuit of substantive goals (such as full employment, export competitiveness, energy self-sufficiency) (Evans, Rueschemeyer and Skocpol 1988; White 1988; Delacroix and Ragin 1981). Often with the support of powerful government institutions, the state actively intervenes in society and the market – at times in an authoritarian manner. One might wonder whether the environmental toll of rapid economic growth and concurrent industrialization in NICs is not caused by the developmental nature of the state itself, and if so, whether changes in the governance style of the state might not lead to a better environmental performance of the industrial and business sector.

Opposed to the more optimistic overtones of the proponents of the greening of industries, we find those who posit that greening is little more than "greenwash" or a "zero-sum" game. In the eyes of more radical environmental activists, measures such as the use of voluntary, consensual approaches in the greening of industries run the danger of sanctioning "business-as-usual." Due to their profit-oriented nature, businesses tend to avoid the implementation of costly environmental measures. Asking them to do this voluntarily, rather than through (governmental or civil society) pressure is merely leaving the fox to watch the geese, and will lead to the application of a veneer of greenwash rather than a fundamental greening. In this respect, it is noteworthy that a recent report by the Dutch Environmental Inspection Agency concluded that the environmental reporting by a large proportion of companies had serious shortcomings.[3] For these reasons, Friends of the Earth, a transnational NGO, dubbed the voluntary approach "undemocratic, ineffective and inherently weak." Greening without pressure and confrontation is seen as the cooptation of adversarial activist politics and avoids the discussion about fundamental changes in production and consumption. As Christoff (1996) stated, the debate about greening industries is rhetoric that in itself has become "a strategy of political accommodation of the radical environmentalist critique of the 1970s."

For analytical clarity, with regard to the greening of industries one needs to distinguish between the discussion about corporate environmental behaviour of transnational companies (TNCs) versus that of domestic enterprises in developing nations. Concerning the former, a substantive part of the debate tends to focus on the globalizing forces behind and the environmental impact of flows of foreign direct investments (FDI) from industrialized economies in the North towards industrializing economies in the South. Antiglobalists would argue that flows of FDI are predominantly propelled by a corporate quest to find havens of low environmental standards, weak pollution control enforcement and cheap labour costs.[4] Others counter this argument on the grounds that corporate environmental regulations and parent company pressure will lead to the greening of branch companies overseas and might even have a positive effect on the environmental performance of industries in developing economies. In this volume we will not deal with the FDI flows from North to South, but

[3] Based on an inspection of 251 Dutch companies. Report by the Milieu-inspectie, cited in NRC, "Milieuverslag bedrijf 'schiet ernstig tekort'" (Companies' Environmental Report 'highly inadequate'), *NRC Handelsblad*, 8 May 2001, p. 4.

[4] See the various contributions in Utting, Peter (ed.), (2002), *The Greening of Business in Developing Countries: Rhetoric, Reality and Prospects*, ZED Publications, London.

address a different discussion: the greening of domestic industries in developing nations and the role of novel environmental governance structures in achieving this.

In this respect, we need to consider the nature of the state in relation to the socioeconomic context in which it operates. To varying degrees the initial period of rapid economic growth in China, Vietnam, Singapore and Taiwan, have been achieved through centralized, interventionist state structures with a commandist fashion of policy implementation. At the same time, the economic growth in the NICs – and particularly so in China and Vietnam today – has also led to the "co-existence of the First and Third Worlds" with their ensuing social cleavages and economic frictions. Depending on the level of socioeconomic development, the East-Asian NICs are confronted with developmental dilemmas ranging from food security, rural poverty, booming urbanization, and environmental degradation. In the face of major development problems that still await solutions, an interventionist, developmental state that regulates society through command-and-control might be the resultant rather than the cause of a particular socioeconomic constellation. This implies that the emergence of new governance structures are equally propelled and affected by that constellation.

A similar line of argument has been put forward in the analysis of the interaction between economic growth and political reform in the Asian tiger economies. Thompson (1996) remarked that "the success of economic development itself in these Asia-Pacific countries was also crucial to the stability of authoritarianism (...). Successful export drives help account for the 'lateness' of democratisation in the Asia-Pacific." In this light, one should be cautious in advancing the idea that the democratic degree of the state, and its environmental policies for that matter, stand on a straightforward one-to-one relation with the effectiveness of industrial pollution control. Such notions open up avenues for a mere imposition of policy formats with scant regard for local variations, and run the risk of demonstrating the "naïveté and analytical shallowness of the current orthodoxy in assuming that democracy can somehow be made to happen, but also that it will work on a sustained basis and that it will promote growth" as Leftwich (1996) so aptly phrased it. Against this backdrop, we can ask ourselves whether the implementation of policy measures that entail decentralization, voluntary coregulation, participatory and market-based approaches, might not work counterproductive in the East-Asian NICs. More importantly, one might question ideas of "leapfrogging development" so reminiscent of the development rhetoric of the 1960s and 1970s, because it is uncertain to what degree "soft" and market-based state regulation on the premises of a democratic polity would work within a newly industrializing context and a different political system.

In sum, rather than arguing that decentralization, public participation, voluntary coregulation and other measures are eventually a *conditio sine qua non* for improved industrial pollution control, do we need to assess the feasibility and desirability of such measures in the light of the past and existing socio-economic and political conditions of developing nations. The majority of the contributions in this volume demonstrate that changes in industrial pollution control under the rubric of industrial greening actually run counter to the developing context in NICs. It is not the aim of this book to provide a comprehensive assessment of different instruments in industrial pollution control and environmental management. Rather it is our aim to identify and explain some of the similarities and differences that exist between the Chinese, Vietnamese, Taiwanese and Singaporean cases, and gauge whether, and to what degree, we might extrapolate these results to a more general level of analysis. In various ways, the contributions touch on questions, such as the role and the reach of developmental states in the governance of industries; the fallacy in assuming "leapfrogging" development; the feasibility of market reforms, adoption of green technologies, and changes in environmental governance styles in a newly industrializing context. These questions critically depend on the political structure and level of state capacity in the NICs. For these reasons, we need to differentiate between the different contexts in the four NICs.

3. Relevance of the Cases

Scholars have generally attributed the economic successes of the NICs to factors that typically included:

- The role of state intervention – the "developmental state" with its support for the domestic industrial sector through the establishment of strong planning bodies, the use of subsidies, tariffs, pricing policies and "juggled" exchange rates;
- The structure of industrial and corporate networks – e.g., large state corporations such as the Japanese *keiretsu* and Korean *chaebol*, as well as family networks, such as the Chinese *qiaoxiang* ties;
- The strategy for industrialization – i.e. the shift from labor-intensive, import substitution industrialization to capital-intensive, export oriented industrialization;
- The level of state autonomy ("embedded autonomy") – the degree to which the state can formulate and enact policies independent from the vested interests of landlords, the industrial bourgeoisie, the working class, and the military, as well as bilateral and multilateral institutions (e.g., the World Bank, IMF and United Nations).

The Chinese, Vietnamese, Taiwanese and Singaporean economies share a rapid economic development with growth rates of respectively 10.1, 6.0 and 7.7 percent average annual growth in GDP during 1991-2000 (World Resources Institute 2003). Furthermore, to various extents the aforementioned factors also determine the industrial dynamics in the four NICs on which the contributions in this volume draw: China, Vietnam, Taiwan and Singapore. However, there are also important divergences that stand out among the four NICs in terms of their governance structure, the trends in industrialization, and their environmental record.

The differences in the nature of the governance structures between Vietnam, Taiwan, Singapore and China need ample consideration when analyzing industrial greening. With its semi-autocratic polity in which civil liberties and the freedom of speech are partly limited, Singapore is posited at a middle position of the governance spectrum. At one end of the spectrum, we find the Chinese and Vietnamese one-party states scoring low on the polity indices of Freedom House, whereas Taiwan occupies the other end, featuring a full-fledged liberal democratic system, a free press, and a high level of civil liberties (see Table 1).

Due to its democratic governance structure, Taiwan has a relatively forceful environmental movement that on several occasions has successfully prompted businesses and the government into environmental action. In China, Vietnam and to a lesser extent,

Singapore, environmentalism is still largely constrained within the social and political limits charted out by a (semi)authoritarian state (Ho and Edmonds 2008).

Table 1. Overview of selected governance indices in China, Vietnam, Singapore and Taiwan[5]

Governance index	Region			
	China	Singapore	Taiwan	Vietnam
Level of political freedom (free, partly free and not free)	Not free	Partly free	Free	Not free
Polity index in 2000 (-10 = fully autocratic; 10 = fully democratic)	-7	-2	9	-7
Political rights (1 = most free; 7 = least free)	7	5	1	7
Civil Liberties (1 = most free; 7 = least free)	6	5	2	6
Press freedom (1-30 = free; 31-60 = partly free; 61-100 = not free	80	68	21	82
NGOs per million inhabitants in 2000	2	477	not available	10
Corruption index (10 = least corrupt; 0 = most corrupt	4	9.2	6	3

Source: Freedom House, cited in World Resources Institute, EarthTrends 2003, country profiles, http://earthtrends.wri.org.

In China and Vietnam, there are relatively few nongovernmental and voluntary, civic organizations, whereas confrontational actions and public disclosure through the mass media are difficult (Ho, 2001). Also in Singapore environmental activism has only been burgeoning over recent years, and most NGOs have to liaise closely with the government and industry in order to be effective. In addition, the state closely coordinates activities of environmental groups through the Singapore Environment Council (formerly the National Council for the Environment).

Although all labelled as newly industrializing economies, the four NICs have followed widely diverging development paths. As two of the four Little Tigers, both Singapore and Taiwan are deemed textbook examples of first-generation NICs. Both economies have been early in shifting from import substitution industrialization towards export oriented industrialization. The Taiwanese state has not gone in for state-owned heavy-industry projects, but has instead focused on providing the infrastructure and investment climate conducive for the labor-intensive production of consumer goods by private entrepreneurs. In the 1950s, Taiwan encouraged the growth of import-substituting industries, which resulted in a gradual shift of industrial production away from food processing for export to textiles, rubber and leather goods, bicycles, wood products and other consumer items produced for the domestic market. Yet, by the end of the decade the scope for further expansion of import-substitution by the small and medium-sized enterprises (SMEs) had been exhausted. In

[5] Unless indicated otherwise, data pertains to the year 2001.

reaction, the government changed its industrial policy to stimulating the export of nondurable consumer goods already being produced in Taiwan. Industrial production grew at 20 percent per year during the 1960s with an increased importance for electronic components in the second half of the decade. This trend continued in later decades, with the industrial sector gradually having more capital, knowledge and technology-intensive. Singapore has followed a slightly different path of development.

At the time of independence from the British Empire in 1959, Singapore had inherited a mixed industrial legacy. The industrial sector was small and its productivity low. In 1960 manufacturing was only a mere 11.4 percent of the GDP, while commerce, which was by far the largest sector, accounted for 32 percent. The government opted for a traditional import substitution strategy that sought to promote industrialization as a way of diversifying from Singapore's traditional role as an entrepôt. However, after a short-lived two-year union with Malaya to form Malaysia and the subsequent breakup in 1965, Singapore had little other recourse than to decisively turn to export-oriented, labor-intensive industrialization. Initially, industrial development relied on manufacturing to propel economic growth, with a strong emphasis on electronic and electrical products. It was at this time that the Singaporean government began to formulate a strong state-guided industrialization program implemented through the Economic Development Board. Different from Taiwan, domestic SMEs played a lesser role in export-oriented industrialization. In addition, more than other Asian NICs Singapore has heavily relied on foreign direct investments which gave rise to a virtually complete dominance of TNCs in the manufacturing industry.[6] It was not until the economic recession of the mid-1980s that the Singaporean Economic Development Board attached greater importance to stimulating and upgrading the small and medium-sized domestic industries.

Although industrialization in China and Vietnam can be traced back to the 1950s and 1960s, the industrial takeoff occurred at a much later stage. Not only after the economic reforms were well under way – in China since 1978 to be followed by Vietnam eight years later – was a solid basis laid for a NIC-type of industrialization. Yet, China and Vietnam are no carbon copies of their East Asian neighbours, if not only for the fact that both countries had to shake off a socialist past during which the industrial sector was controlled through soviet-style central planning. The mainstay of the Chinese and Vietnamese economies in the socialist era was formed by large-scale, state-owned industries. By illustration, during 1956-1978 the ratio of the gross industrial output value of state versus collective industries in China typically hovered around 9:1 or 8:2. Bringing the market back into the industrial sector implied the reform of resource-inefficient, state-owned industries that for decades had operated on obsolete technologies, yet, traditionally had also provided employment to a large proportion of the working populace. The economic reforms – popularly called *Gaige Kaifang* in Chinese and *Doi Moi* in Vietnamese – did not so much lead to an industrial takeoff in the urban, state-owned sector, but occurred mostly within collective industries located in the rural and periurban areas. In addition, also the rise of other ownership types of enterprises – i.e. private companies and joint ventures – took place.

[6] During 1980-1990 Singapore received more foreign direct investments in absolute terms than any other less developed country. Singapore Department of Statistics, *Yearbook of Statistics*, (Singapore: Singapore National Printers: various years).

It is important to note that SMEs have played a critical role in the economic growth of the Asian NICs.[7] As we can see from Table 2 below, SMEs in China, Vietnam, Taiwan and Singapore make a significant economic contribution in terms of employment and industrial output value (UNCTAD 2003).

Table 2. Economic contribution of SMEs in selected Asian NICs

Country	Percentage of SMEs of total enterprises	Employment (percentage)	Percentage of industrial output value
China	99	73	60
Taiwan	98	80	36
Singapore	97	58	41
Vietnam	unavailable	56	42

Source: UNCTAD, *Improving the Competitiveness of SMEs through Enhancing Productive Capacity*, TD/B/Com.3/51/Add.1 (Geneva: 31 January 2003), table 2, p. 3 and country profiles; and Taiwan Small and Medium Enterprise Working Group, (2003), "The General Situation of Development of the Small and Medium Enterprises" at http://www.taipei.org/press/smea.htm accessed at 22 November 2003.

When studying the dynamics of industrial pollution control in the NICs, one needs to consider the type and scale of enterprises. For one thing, in contrast to large-scale industries, SMEs are generally short of capital, and as a result, they operate with outdated and polluting machinery, and have limited ability to adopt clean technologies. In addition, due to the low barriers to establish enterprises of this kind, competition between them is fierce. Most SMEs are under strong pressure to produce at the lowest possible price, regardless of the environmental consequences. Lastly, SMEs – in particular during early stages of development – are often found in residential areas, where they lack access to even the most basic sanitation services such as sewers and waste disposal facilities. In combination with badly trained managers with low environmental awareness, it should come as no surprise that SMEs greatly affect the quality of life of the urban and periurban residents.[8] But also in these respects, there are differences among the cases studied here.

In China and Vietnam the small- and medium-sized enterprises are labor-intensive, light industries active in areas such as food processing, garments, footwear and toys. Singapore and Taiwan on the other hand shifted to capital and knowledge-intensive hi-tech industry during the mid 1970s and 1980s. The capital-intensive SMEs in Singapore and Taiwan generally have more ability to acquire new and expensive environmental technologies than their labor-intensive counterparts in China and Vietnam. Moreover, with regard to the relocation of polluting industries to industrial parks and even environmental parks that employ principles of industrial ecology, Singapore and Taiwan are running ahead.[9] The situation in China and

[7] These include Bangladesh, China, Hong Kong, India, Indonesia, Japan, Malaysia, Nepal, Pakistan, the Philippines, the Republic of Korea, Singapore, Sri Lanka, Thailand, and Vietnam.

[8] For more information, see also Arthur P.J. Mol and Joost C.L. van Buuren (eds.), (2003), *Greening Industrialization in Asian Transitional Economies*. Lanham, Maryland: Lexington Books.

[9] For instance, since fall 2002, the Taiwan Environmental Protection Agency has committed a total budget of NT$ 5 billion for the development of Environmental Science and Technology Parks. In total, three such parks – two in Kaohsiung and Hualian counties, and one which location is still undecided – will be established, while plans for a fourth park are still pending the approval by the relevant authorities. See Environmental Protection

Vietnam is still serious with a substantial proportion of the SMEs lacking basic access to waste treatment facilities. In Vietnam there are 62 industrial zones, of which only 22 have completed infrastructural construction and just five have central effluent treatment plants in operation (UNEP 2002). The situation in China is slightly better, although a substantial proportion of the SMEs still lack basic access to waste treatment facilities. A recent positive trend is the establishment of demonstration environmental industrial parks in paper and sugar manufacturing, such as the State Guigang Ecological Industrial Park, the Nanhai Ecological Industrial Park, and the Xinjiang Shihezi Ecological Industrial Park.

Industrialization in China, Vietnam, Taiwan and, less so in Singapore, has been achieved at the expense of the natural environment (Ho and Vermeer 2006). As newly industrializing economies there are marked parallels in problems of air and water pollution, and the explosive growth in the generation of solid waste. However, differences in industrialization among the four NICs have also caused diverging and, at times, contradictory trends in environmental performance. On the one hand, earlier industrializing economies such as Singapore and Taiwan have been able to effectively deal with some of their initial environmental ills. For instance, whereas China and Vietnam are still struggling to control and bring down the consumption of ozone-depleting Chlorofluorocarbons (CFCs),[10] Taiwan has already successfully phased out CFC production and has reduced consumption to zero since January 1996. Also Singapore has made great strides in the control of CFC consumption, and has brought CFC consumption down from 679 kg CFC per 1,000 persons in 1989 to a mere 1 kg in 2002.[11]

Furthermore, in contrast to Taiwan the Singaporean government has been relatively swift in avoiding a "pollute-first, clean-up-later" type of industrialization strategy, for which it has gained the reputation of a "green garden city." For example, the annual average levels of air pollution since they were first monitored in 1975 have always met the international standards on air quality set by the World Health Organization and the United States Environmental Protection Agency. Singapore claims it has effectively combated environmental problems typical of fast-growing cities in developing countries, such as infrastructure inefficiencies, traffic congestion, industrial pollution, incompatible land uses, and urban sprawl. Observers have pointed to a set of factors that might explain Singapore's environmental successes: its interventionist state headed by a committed, non-corrupt leadership since independence in 1959; its literate, well-informed, and law-abiding populace; and the small size of the nation and its one-level government who facilitated efficient communication and coordination. However, some of these explanatory factors raise questions about the feasibility of repeating a "Singaporean model" in other NICs.

Even more important is it to note that the very *nature* of environmental pressure has shifted during the process of urbanization and industrialization. This shift becomes obvious in the differences in energy-intensity and carbon dioxide emission per capita among the NICs (see table 3). Seen in this light, the environmental performance of the Singaporean "green garden city" is actually far worse than Taiwan's. The high-level of industrialization of Taiwan and Singapore is reflected in a level of energy-intensity equal or even higher to that of

Agency (2003), "Bidding Begins for Third Environmental Science and Technology Park", EPM, Vol. 6, No. 1, January, at www.epa.gov.tw.

[10] For instance, in China, the level of CFC consumption per kg per 1,000 persons was 30,621 in 2002, which after a strong increase over the 1990s is basically back at the level of 1989 (34,783).

[11] Data from UNEP Regional Resource Centre for Asia and the Pacific, available at http://www.rrcap.unep.org.

industrialized nations in Europe and the United States. With a per capita energy consumption of 5,791 kg oil equivalent, Singapore is almost twice that of the European average, and over three times the world average. China, on the other hand, has an energy consumption per capita close to the Asian average, while Vietnam only has half of that. And with a doubling of the number of aircos and almost a tripling of the number of cars owned per household in just four recent years (from 1999-2002), it can be expected that China's energy consumption will soar in the decades to come. The recent power cuts in the coastal zones are a bleak testimony to this.

Table 3. Selected energy and CO_2 emission indicators

	China	Taiwan	Singapore	Asia (excl. Middle East)	Europe	United States	World
Total energy consumption (1,000 metric tons of oil equivalent)	1,088,349	79,925	22,693	2,919,333	2,559,701	2,269,985	9,702,786
Energy consumption per capita (per kg oil equivalent)	861	3,618	5,791	867	3.516	8.095	1.623

Source: World Resources Institute, *Earthtrends 2003*, country profiles, section climate/atmosphere and energy/resources at http://earthtrends.wri.org; World Resource Institute (2003), *World Resources 2002-2003*. Washington D.C.: WRI.

Considering the carbon dioxide emissions per capita, we see that China and Vietnam feature relatively low emission levels characteristic of industrial latecomers (the higher Chinese level is mostly caused by the heavy reliance on coal to fuel the economy). Contrarily, the high level of urbanization and industrialization in Taiwan and Singapore have already led to emission levels that by far surpass those of Western Europe (see Figure 1). Whereas some would argue that industrial greening in the long term might offer prospects for the delinking of economic growth with increasing environmental pressure,[12] the Taiwanese and Singaporean trends could point to a potential relinking as not the scale, but the character of environmental pressure changes over time. According to Opschoor (2000), the gradual environmental reforms as proposed for greening industries are insufficient:

> [M]ore elaborate empirical investigations cast some shadows over these bright prospects. (...) For several countries environmental pressure appears to have been relinked with the environment since the mid-1980s. (...) The message of recent empirical analysis may be that endogenous delinking does not appear as a process which is stable or persistent under conditions of sustained economic growth. Sustained growth is not necessarily sustainable.

Similar conclusions were reached by the European Environmental Agency, which found that the environmental measures and technological innovations of the past decades had been

[12] As is described in the so-called environmental Kuznets curve.

offset by the sheer increase in the scale of production and consumption. Although interesting in its own right, the "de- and relinking" debate will not be the main focus of this volume, but the feasibility and desirability of introducing novel environmental governance structures against a developing and newly industrializing backdrop.

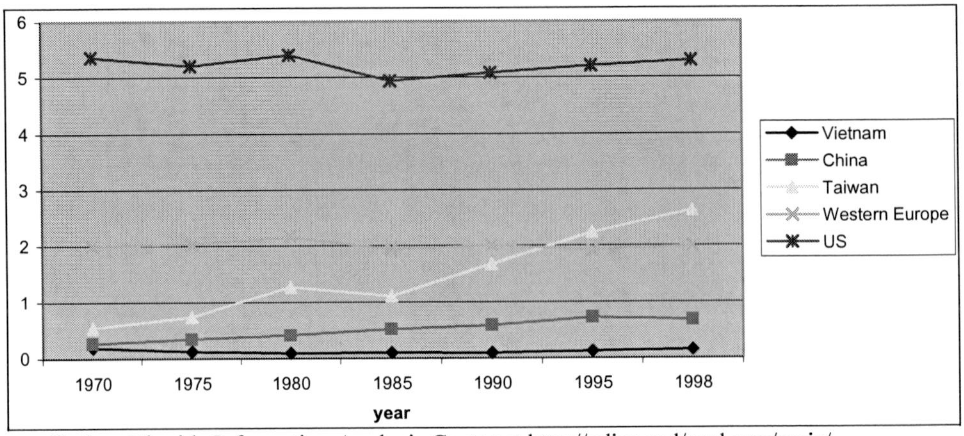

Source: Carbon Dioxide Information Analysis Center at http://cdiac.esd/ornl.gov/emis/

Figure 1. CO2 emissions per capita over 1970-1998.

4. Opportunities and Constraints in Leapfrogging

The Contributions

In the environmental debate it is often suggested that developing countries do not necessarily need to go through the stages of "dirty development" that typified industrial growth of developed nations.[13] By installing the right environmentally friendly technologies, and establishing the appropriate legal and political framework, developing countries might leapfrog over the past mistakes of today's industrialized economies. Such perspectives are particularly appealing for NICs that find themselves in the midst of processes of rapid industrialization and environmental degradation. However, Perkins aptly warns us that:

> These are bold claims as they suggest that many of the environmental problems commonly associated with urban industrialization may be a thing of the past. The question is whether leapfrogging is a realistic and achievable goal. (...) [H]owever, current approaches may be overoptimistic. Meeting the objectives of clean development in industrializing countries is likely to prove considerably more complex and challenging than much of the existing literature would lead us to believe.
>
> (Richard Perkins 2003)

[13] See, for example, J. Goldemberg, (1998), "Leapfrog energy technologies", *Energy Policy*, Vol. 26, No. 10, pp. 729-42.

In fact, the various contributions in this volume are testimony to the complexity of achieving, let alone, leapfrogging towards clean development.

Since decollectivization in the mid-1980s, the People's Republic of China has made great strides in legal reforms. In 2002 China promulgated the new Cleaner Production Promotion Law. The law is of vital importance as it sets forth the national framework for the greening of industries – although its scope extends to the realms of agriculture and the service sector as well. The law signifies a break with the past during which the responsibility for environmental governance was put squarely on the shoulders of the state that had to regulate industries through administrative and punitive measures. However, the proclamation of the Cleaner Production Promotion Law intends to effect a shift in governance style towards the common belief that cleaner production can only be achieved *within* and by the industrial sector itself, rather than being imposed from above. This implies a renewed role for the state: one of facilitating, promoting and enabling the transition from curative and end-of-pipe practices to preventive cleaner production.

The first contribution by Liu and Mol provides an in depth analysis of the political debates involved in the drafting of the Cleaner Production Promotion Law in China, and examines the possible consequences on the industrial sector. In order to effect so-called "win-win" scenarios – in which economic competitiveness can be combined with environmental quality – the law stresses government support, economic incentives and voluntary actions by enterprises, instead of state regulation and intervention. In sum, for the greening of industries China needs to rely on a facilitating rather than a policing state. This chapter is followed by a contribution by Cushing who investigates the status of the ISO 14001 Environmental Management System (EMS) standard in China and analyzes key factors influencing its adoption by Chinese enterprises. It applies lessons learned from studying the implementation of China's mandatory environmental regulations to assess the potential for using this voluntary standard to promote improved environmental performance. Her findings indicate that the number of certified enterprises is growing exponentially and that factors influencing adoption include international trade, transnational corporate policy, government-sponsored environmental projects, the potential for regulatory, economic, and environmental benefits, and the interest of top company management.

Apart from the appropriate institutional and legal environment, another critical factor in the greening of industries is, of course, the availability of environmental technology and services, as well as the sector that could provide these: the environmental industry. The contribution by Liu *et al.* reviews several fundamental constraints in the development of China's environmental industry. At the national and regional levels the environmental industry is impeded by barriers relating to the market demand and supply, whereas the hierarchical bureaucratic structure and lack of appropriate governmental regulations limits the opportunities for technological innovation. In absolute terms of firm numbers and sales, the environmental industry in China might look impressive, yet comparatively, it still lags far behind that of the United States and the European Union. There are several factors that have a negative impact on the overall development of the environmental industry. First, environmental industries are generally scattered, small-scale enterprises with low economies-of-scale, insufficient knowledge and capital, and weak innovative capabilities. Second, the quality of most environmental equipment, products and services still leaves a lot to be desired, and only a very small number of enterprises manages to meet current international standards. And finally, market imperfections and regional protectionism significantly undermine the innovative and R&D capacity of environmental industries.

Moving from one to another socialist economy in transition, the contribution by Khoa and Ho looks at the Vietnamese efforts in industrial pollution control. Industrial pollution problems in the capital HoChiMinh City are becoming increasingly serious. With more than 28,000 small- and medium-scale industries and over 700 large-scale industries that are mainly situated in residential areas environmental protection is posing a complex challenge to local authorities. At present, a program for the relocation of polluting industries into industrial areas and neighborhoods is emerging as a focus of the city authority's concerns. The program includes industry and pollution inventories, a selection of priority industries for relocation and social and financial support mechanisms. In this chapter the benefits of relocation are evaluated, as well as the practical constraints in the implementation trajectory. Major barriers for successful relocation are related to the absence of land in industrial zones, lacking resources to finance the relocation process and the typical characteristics of the city's (small- and medium-sized) enterprises.

It might be insightful to liken the experiences in industrial greening of industrializing latecomers, such as China and Vietnam, to that of Taiwan – a first-generation industrializing "tiger economy" *pur sang*. Over the past decades, Taiwan has gradually emerged as a top producer of information technology (IT). The IT sector has the image of an advanced, highly capital and knowledge-intensive industry. Along with this image, IT is often regarded as a clean industry: wrongfully, as the following two contributions will demonstrate. Although both chapters examine the greening of the IT industry, they are in various ways complementary. The chapter by Tu examines the specific features of the IT industry itself, and determines to what extent these affect the environmental governance of this sector. Tu argues that the characteristics of IT – in particular, its network-based, specialized production; regional competitiveness; and rapidly changing nature – while providing ample economic opportunities, at the same time pose great environmental challenges. In the Taiwanese context, these challenges are further amplified: the IT sector is heavily nurtured and supported by the developmental state itself which adds to its dominance.

The contribution by Yang takes a slightly different perspective than the previous chapter. Yang's chapter takes a bird's eye view of environmental reform in the political, social and economic spheres – with the IT sector as a case study. Whereas Tu paints a more bleak picture of industrial greening of IT, Yang also points to opportunities for environmental reform. The factors that influence and support the Taiwanese government's efforts in environmental protection include democratization, growing public awareness about environmental issues, international competitiveness, gaining first-mover advantage, and export opportunities for green technology to less developed countries. Although Taiwan's environmental activism is relatively underdeveloped in comparison with other industrialized nations, the prevalence of a "sustainable development" discourse, and the adoption of environmental management and other technical and policy innovations seem to suggest that it is catching up with its counterparts in the West. On the other hand, however, Yang also admits that the results of the state's dematerialization and industrial greening efforts are still too early to judge.

The final chapter of the book by Lang reviews how a land-scarce city-state, such as Singapore, is trying to achieve its goals of zero landfill and zero waste through the greening of industry. The main challenges Singapore confronts in its solid waste management are an increasing volume of waste generated, a shortage of land for landfills, and escalating costs of incineration plants. To green its industries, there has been a coordinated effort to develop a recycling industry and to initiate public-private partnerships that will advance environmental technology. Strategic tie-ups in steel, construction, waste incineration, and the food retail industry illustrate the progress that has been made. Unfortunately, many studies have found that the environmental performance of the construction industry in Singapore leaves much to be desired. The industry has a low level of awareness of environmental auditing and a general reluctance to take action other than those needed to meet regulatory requirements. It is not ready for systematic environmental management. Also the food industry has a long way to go towards greening, as it features a dismal rate of recycling with no one willing to bear the costs of recycling food waste. In addition, the industry provides minimal disclosure on environmental policy and environmental benefits of products or processes, and announcement on ISO 14001 certification.

Coming back to the quote with which this section started: how realistic and achievable is the claim that NICs are uniquely situated in a position in which they could actually opt for a cleaner development path than that of the industrialized economies? Judging from the contributions here we can safely conclude that the newly industrializing context – despite its many economic opportunities – teaches us to tone down high-strung expectations of leapfrogging. Certainly, there are differences in the Chinese, Vietnamese, Singaporean and Taiwanese cases in terms of their environmental governance and performance. But what can also be deduced from these cases is that there are also structural factors at work that impede opportunities for leapfrogged development. These typically include political and institutional constraints, market imperfections, and the nature of the developmental state. Moreover, expectations of leapfrogging also exhibit an incomplete understanding of the industrial dynamics in latecomer industrializing economies. For example, the industrial landscape in China and Vietnam is dominated by small- and medium-scale industries that operate on outmoded and highly polluting technologies. For such enterprises, the substantial capital investments necessary to install environmental technologies might not be within easy reach. Therefore, developing and newly industrializing economies frequently find themselves in the situation of being "locked into" a dirty development path. The problem arises because once polluting technologies are installed, the only way out of a "lock-in" would be to retrofit "end-of-pipe" technology or, even more costly, altogether replacing the machinery with cleaner production devices that target the entire chain of production.

Pursuing a conscious strategy of leapfrogging places high demands on public policy. Typically, the discussion in this regard has also focused on a redefinition of the state's role. The regulatory, interventionist strategies of the developmental state should make place for a more facilitative approach that allows for stronger self-regulation by industries; the adoption of preventive cleaner production technologies rather than end-of-pipe facilities; and a greater role reserved for nongovernmental actors, such as green NGOs and pressure groups. It needs no mentioning that this would also require a reconfiguration of public policy: policy-making should proactively embark on a leapfrogging development path, rather than the common, reactive approach of slow and incremental "muddling through." In other words, the state needs to define specific targets for leapfrogging (which pollutants need to be brought down and at which level?); it must prioritize certain industrial sectors for cleaner development (which sectors should be singled out with an eye to cost efficiency, the potential capacity and degree of leapfrogging, and the possibility for long term environmental improvement?); and finally, the state should stimulate industries' capabilities for leapfrogging by creating an appropriate political, legal and market environment, as well as by promoting strategic partnerships between key stakeholders.

The question is whether this reconfiguration is feasible in industrialized economies, let alone, in newly industrializing economies, such as in China and Vietnam, that still face major and basic development problems. Even in first generation NICs, like Taiwan and the "green garden city" of Singapore, where the control of industrial pollution has been fairly successful, self-regulation and the implementation of ISO 140001 has met with considerable resistance from industries. Furthermore, although not directly touched upon by the contributions in this volume, the shift in the *nature* of environmental pressure as industrialization progresses is an issue that is not to be neglected. In different ways, the contributions collected here are evidence of the arduous and, at times, seemingly insurmountable struggle towards the greening of industries.

References

Adams, W.M., (2001), *Green Development: Environment and Sustainability in the Third World*. Second Edition. London and New York: Routledge.

China Environment Yearbook, (2002), Beijing: China Environmental Yearbook Press, p. 263.
China National Bureau of Statistics, *China Statistical Yearbook 2003*, (Beijing: China Statistics Press: 2003), p. 352.

Christoff, P., (1996), "Ecological Modernisation, Ecological Modernities," *Environmental Politics*, No. 5, pp. 476-99.

Delacroix, J. and C.C. Ragin, (1981), 'Structural Blockage: A Cross-national Study of Economic Dependency, State Efficacy, and Underdevelopment,' *American Journal of Sociology* 86, pp. 1311-47.

European Environmental Agency, (2003), Europe's Environment: The Third Assessment. Copenhagen: European Environmental Agency, pp. 7, 28.

Evans, Peter B. Dietrich Rueschemeyer and Theda Skocpol (eds.), *Bringing the State Back In*, Cambridge: Cambridge University Press.

--- (1995), Embedded Autonomy: States and Industrial Transformation. Princeton: Princeton University Press.

Friends of the Earth, (2002), "*The Voluntary Approach,*" FON Briefing, August.

Hin, David Ho Kim, Robert The Yoke Chong, Tham Kwok Wai and Clive Briffett, "The Greening of Singapore's National Estate", *Habitat International*, Vol. 21, No. 1, 1997, pp. 107-121.

Ho, Peter, and Richard L. Edmonds (2008), *China's Embedded Activism: Opportunities and Constraints of a Social Movement*, London: Routledge.

Ho, Peter and Eduard B. Vermeer (2006), *China's Limits to Growth: Greening State and Society*, Oxford: Blackwell Publishers.

Ho, Peter. "Greening without Conflict? Environmentalism, Green NGOs and Civil Society in China", *Development and Change*, Vol. 32, 2001, pp. 893-921.

Holme, R. and P. Watts (2000), *Corporate Social Responsibility: Making Good Business Sense*, World Business Council for Sustainable Development, Geneva.

Huff, W.G., "The Developmental State, Government and Singapore's Economic Development since 1960," *World Development*, Vol. 23, No. 8, 1995, pp. 1421-1438.

Jenkins, R. 1991. "The political economy of industrialization: a comparison of Latin American and East Asian newly industrializing countries," *Development and Change*, Vol. 22: 197-231.

Kalland, Arne and Gerard Persoon, (eds.), (1998), *Environmental Movements in Asia*, Richmond, Surrey: Curzon.

Leftwich, Adrian (ed.), (1996), Democracy and Development: Theory and Practice, Cambridge, Polity Press.

National Bureau of Statistics, (1999), Comprehensive Statistics over 50 Years of China, Beijing, China Statistics Press, p. 36.

Opschoor, J.B., (2000), "Industrial Metabolism, Economic Growth and Institutional Change," in Michael Redclift and Graham Woodgate (eds.), *The International Handbook of Environmental Sociology*. Cheltenham: Edward Elgar, pp. 280-281.

Perkins, Richard, "Environmental Leapfrogging in Developing Countries: A critical assessment and reconstruction," *Natural Resources Forum*, Vol. 27, 2003, pp. 177-188.

Porter, G. and J.W. Brown, (1996), *Global Environmental Politics*. Boulder Colorado: Westview Press.

Rip, A. and R. Kemp, (1998), "Technological Change," in: S. Rayner and E.L. Malone (eds.), *Human Choice and Climate Change: Resources and Technology*. Columbus: Batelle Press.

Seik, Foo Tuan. "Urban Environmental Policy: The Use of Regulatory and Economic Instruments in Singapore," *Habitat International*, Vol. 20, No. 1, 1996, pp. 5-22.

Thompson, Mark R. (1996), "Late Industrialisers, Late democratisers: Developmental states in the Asia-Pacifi," *Third World Quarterly*, Vol. 17, No. 4, pp. 625-647.

UNCTAD, *Improving the Competitiveness of SMEs through Enhancing Productive Capacity*, TD/B/Com.3/51/Add.1 (Geneva: 31 January 2003), table 2, p. 3 and country profiles.

UNEP (2002), State of the Environment in Vietnam 2001: Environmental Concerns, at www.rrcap.unep.org/reports/soe/vietnam.

White, Gordon (ed.) (1988), *Developmental States in East Asia*. London: MacMillan.

World Resources Institute, *Earthtrends 2003*, country profiles, section economic indicators at http://earthtrends.wri.org.

In: Leapfrogging Development in Emerging Asia
Editor: Peter Ho, pp. 19-38
ISBN: 978-1-60456-410-5

Chapter 2

THE MAKING AND IMPLEMENTATION OF CHINA'S CLEANER PRODUCTION PROMOTION LAW

Arthur P. J. Mol[1] and Liu Ying[2]
[1]Department of Social Sciences, Wageningen University, Wageningen, The Netherlands
[2]Environmental Protection and Resources Conservation Committee National People's Congress, China

1. INTRODUCTION: CLEANER PRODUCTION VIA LAW-MAKING

The introduction has set out the argument whether Asia's emerging economies can leapfrog the "dirty" stages of development by implementing principles of industrial greening and introducing clean technologies. In this chapter, we will narrow down on the legal aspects of greening in China with particular reference to the codification of principles of cleaner production. China began to establish an environmental regulatory system for industrial environmental management and natural resource conservation from 1979 onwards. Following the promulgation of the state Environmental Protection Law (in draft version) in 1979 (revised in 1989), China started to systematically develop an environmental regulatory system. In 1984 environmental protection was defined as a national basic policy and key principles for environmental protection in China were proposed, which include "prevention over control," "the polluter-pays-principle" (already introduced in the 1979 environmental law), and "strengthening environmental management."[1] Subsequently, a national regulatory framework was formulated, composed of a series of environmental laws (on all the major

[1] Notwithstanding the early introduction of the prevention principle in Chinese environmental law, the practice of environmental management has been strongly focused on pollution control via end-of-pipe measures and a command-and-control approach, rather than prevention, throughout the years. Only marginally environmental practices have touched upon more preventive ideas, i.e., industrial wastewater and solid waste recycling and reuse.

environmental segments, starting with marine protection and water in 1982 and 1984), executive regulations, standards and measures.[2]

After the 1992 UN summit China devoted herself to a sustainable development strategy including a preventive approach. In China's Agenda 21 (published in 1994) cleaner production was put forward more strongly. From then onwards cleaner production was introduced into China. During the 1990s, the cleaner production concept has been put into practice in China, in the beginning especially via development aid projects (Shi, 2003).[3] The advancement of cleaner production practices in China can be distinguished into two phases. The first phase, from 1992 to 1997, focused on the introduction of the cleaner production methodology, the training of personnel and the implementation of demonstration projects at the enterprise level. While significant experience had been gathered in that period, it became also clear that the Chinese cleaner production program faced several weaknesses and was declining in terms of scale and impact (Zhang, 2003). According to Shi (2003, pp. 68-74) these weaknesses related to:

- Lack of a clear definition of the western concept of cleaner production, together with cultural misunderstandings;
- Incompatibility with existing environmental policies (preferring end-of-pipe treatments) and strong divergence of institutional efforts;
- Lack of leadership within the industrial community and the industrial departments in advancing cleaner production, as only SEPA and EPBs triggered the idea;
- Lack of related integrative policies and concepts, such as life cycle analysis, design for the environment, industrial ecology;
- Especially toward SMEs there was a too strong focus on sticks and regulative policies and too little attention on supportive and encouraging measures;
- Limited experience in China with corporate environmental management, which is crucial for institutionalizing cleaner production in firms.

During this first phase it became increasingly clear that only with appropriate policies and enabling measures in place, further implementation and institutionalization of cleaner production is possible, beyond demonstration projects installed with development aid money. So the second phase, since 1997, focuses strongly on cleaner production policy studies and policy-making on cleaner production at the level of the government, culminating in the 2002 Law on the promotion of cleaner production (ADB, 2001; Zhang and Chen, 2001).[4] The National People's Congress of China began the cleaner production law-making in 1998 and it

---

[2] At the national level China has now some 20 environmental laws adopted by the National People's Congress, some 140 executive regulations issued by the State Council, and a series of sectoral regulations and environmental standards by the State Environmental Protection Agency (SEPA).

[3] According to UNEP cleaner production is "the continuous application of an integrated preventive environmental strategy applied to processes, products and services to increase overall efficiency and reduce risks to humans and the environment."

[4] Four major policy studies on CP have been carried out to feed the CP law-making process: a case study project by the China Council on International Cooperation on Environment and Development on Taiyuan City (see below); a set of policy studies funded by the Canadian International Development Agency, a study by the Sian Development Bank in the context of their Technical Assistance Cluster Project for the Promotion of Clean Technologies, and a CP project in the EU-China Liaoning Integrated Environmental Programme (ADB, 2001; Zhang and Chen, 2001). The integral official English version of this law can be found at http://www.chinacp.com/.

took them nearly four years to adopt it rather smoothly in June 2002 as we will see below. From the first of January 2003 the Cleaner Production Promotion Law came into effect, launching a new, third phase in Chinese cleaner production activities.

By 2003 in a total of 24 provinces, autonomous regions and municipalities have launched pilot projects on cleaner production. Through cleaner production auditing and technological innovation, the enterprises have achieved both favorable economic and environmental records. On average, the polluting emissions to water and air were reduced by over 20% (http://chinacp.com). These records and successes should not surprise us, as cleaner production has been experimented and introduced in all developed and most of the industrializing countries, often via special supportive programs, technology development assistance and specific organizations. Most countries report equal environmental successes in cleaner production practices and projects. What is surprising, however, is that only in China policy-makers choose – be it after strong debates – to advance and stimulate cleaner production in industry via a specific legal instrument. All other countries have either set up soft policy instruments for cleaner production promotion (subsidy schemes, technological support programs, or voluntary audit schemes), have used indirect instruments, such as company environmental reporting obligations, or used legal instruments in existing environmental laws to indirectly push for cleaner production. In this chapter we will analyze the backgrounds behind the legalistic push towards cleaner production, the law-making process and the choices made in that, and the specific content of the Cleaner Production Promotion Law. In doing so we aim to shed light on the rational behind a legal instrument for cleaner production promotion in China. Information for this study comes from archive research at the NPC's Environmental Protection and Resources Conservation Committee, interviews with officials and experts involved in the policy making trajectory on cleaner production and literature study.

2. The Debate on a Special Cleaner Production Law

Some current environmental policies, principles and measures, such as permits, and discharge fees, have not been stipulated in the cleaner production law. Not unlike some of the cleaner production protagonists in western countries, some Chinese experts involved in the law making process suggested that these conventional measures on pollution control were barriers for cleaner production application. These cleaner production protagonists argued that in the 1990s conventional environmental protection measures in China lacked stimulating mechanisms for enterprises to conduct further pollution reduction; that most conventional measures limited the creativity, knowledge and technical capabilities of enterprises in pollution control; and that these conventional environmental ideas have become only a burden on enterprises, by hindering profit making. As such environmental protection and pollution control was too often stalemated via economic excuses. Cleaner production, with its promise of win-win, needed to stay away from these kind of conventional measures.

But in analyzing the existing environmental laws before the cleaner production law entered into force one can identify cleaner production ideas and measures in several of these laws (see Appendix 1). Though these 'conventional' environmental laws were not designed for cleaner production, in their implementation they did promote cleaner production in some regions or on some issues (such as energy saving, pollution reduction, toxic prevention). Consequently, it should not be too surprising that some experts and governmental organizations (such as the Supreme Court) questioned the necessity of a special cleaner production law to advance cleaner production implementation, as the legal possibilities were already there in other environmental laws. But a major revision of the existing environmental laws to include cleaner production was believed to be more costly, less effective, more time consuming and more difficult. Many of the experiences and problems of implementing the 'cleaner production' measures of conventional environmental laws have been an important reference and input for designing the new cleaner production law.

The other alternative put forward for a special cleaner production law was national policy. Such a national policy would have the advantage, according to some scholars, of having a larger flexibility, which was always considered essential for cleaner production. The counterargument related to the weaknesses of national policies, especially relevant in a situation where most existing environmental laws pushed towards add-on technologies. In addition, strict governmental responsibilities were seen as an essential precondition for successful cleaner production policies, and a legal instrument could set these responsibilities (Du et al. 2001).

Compared with the cleaner production stipulations in these conventional environmental laws, we can notice some remarkable differences and innovations in the cleaner production law. First, the cleaner production law for the first time makes clear what cleaner production is. This is very important, not in the last place for enterprises that did not know what it is but constantly claimed doing it. Second, conventional environmental laws have only few encouraging articles and lack awarding stipulations. Enterprises conducting cleaner production were hardly affected by these laws. The new cleaner production law has specific articles on incentives, legal liabilities and awards. Finally, the conventional environmental laws lack detailed regulation on cleaner production implementation. The so-called "administrative units," "comprehensive units", "local governments," and especially "relevant units" stipulated in those laws are often not clear. What is more, their responsibilities on implementation and supervision of their work have no legal status and were frequently adjusted. The cleaner production law included these lessons and made clear responsibilities for every relevant governmental department and for governments at all levels. Moreover, their cleaner production responsibilities were legalized. It is the first time that an environmental law so clearly stipulates the different responsibilities of implementation organizations.

3. The Law-Making Process on Cleaner Production

In the People's Republic of China the National People's Congress (NPC) is the highest legislative body. It enacts and amends basic laws concerning criminal offenses, civil affairs, state organs, and other matters. The Environmental Protection and Resources Conservation Committee (EPRCC) of the National People's Congress is one of its nine special committees, which form permanent organizations representing the NPC. When the NPC is in session, the main work of these committees consist of studying, examining and drawing up motions related to the law making activities of the NPC. When the NPC is not in session, these committees work under the direction of the NPC Standing Committee. The tasks of the

EPRCC concentrate on drafting environment and natural resources legislation, amending environment and natural resources laws and supervising the implementation of these laws and amendments. The EPRCC has about 20 members who are all NPC delegates with a serving term of five years.

The standard law making process in China can be divided into two steps: the submitting of law proposal and the adoption of law. The right to submit a proposal for law is regulated by law to the State Council, to the Special committees of the NPC, to more than 15 members of the Standing Committee of NPC, or to more than 30 members of the NPC. When a proposal is adopted by the NPC, the following procedures will be ensued. It starts with the installment of a drafting committee and a working committee. Subsequently the relevant department of the State Council will be entrusted to make a first draft of the law. After that the drafting committee and the working committee carry out research and investigation, often resulting in a revision of the draft. Subsequently, the revised draft is submitted to the plenary conference of the special committee for further revision and approval. Then the agreed upon draft is sent to the State Council and to local and regional People's Congresses for discussion, opinions and comments, often leading to further revisions. That version is finally submitted to the plenary session of the special committee, which decides on the final draft law that will be submitted to the Standing Committee. Normally the draft law adoption process goes through three readings by the Standing Committee.

The drafting committee for this law on cleaner production was organized in 1998. But in fact law-making on cleaner production commenced earlier in Taiyuan Municipality, where in 1997 efforts started to formulate local regulations for cleaner production, with the support of SETC, SEPA and UNEP.[5] In October 1999 a local law was adopted by the 19th session of the 10th People's Congress of Taiyuan Municipality, and in November 1999 it was approved by the Shanxi Provincial People's Congress.[6] The national law-making process profited much from the experiences and discussions in this local law-making process (and also from the "Instructions to Speed up CP" from Jiangsu Province; Zhang and Chen, 2001). Discussions on the legal versus the voluntary character, on the definition of cleaner production and on the tasks and responsibilities of different state agencies already played a major role in the Taiyuan cleaner production law. Jiangsu province, with its well-developed market economy, especially focused on the use of economic instruments and incentives to promote cleaner production.

On the national level four phases can be distinguished in the law-making process on cleaner production. In the first phase (May 1998-August 1999) the drafting committee of the EPRCC entrusted the State Economic and Trade Commission (SETC), and surprisingly not the State Environmental Protection Agency (SEPA), with inception research into cleaner production regulation and with the drafting of the cleaner production law.[7] The initial

[5] Taiyuan was the first so-called demonstration city for cleaner production, later to be followed by nine others (Beijing, Shanghai, Tianjing, Chongquing, Jinan, Shenyang, Kunming, Lanzhou and Fuyang), as well as five industrial sectors. The Cleaner Production Demonstration Plan outlines the objectives and activities for these cities and sectors.

[6] During this law-making process 15 seminars were held, over 600 experts were consulted and 17 drafts were formulated and discussed. The law finally has 5 chapters and 39 Articles (Liu, 2003).

[7] Mr. Li Meng, vice chairman of EPRCC was appointed as the chair of the drafting committee, completed with four additional members also coming from the EPRCC; no doubt the better access of SETC to the industrial community played a role in giving this state commission - rather than SEPA - a leading role, not only in the drafting process but also in the implementation of the law. Most directories, lists, implementation regulations

investigations by SETC were focused on the definition and meaning of cleaner production, the relations of a new cleaner production law with other existing laws, the questions related to designing cleaner production encouragement and punishment measures in the law, and the question regarding the executive body of the law. SETC tried to follow new ideas on state-market relations in developing the law, known as "government guide, market push, enterprise conduct, the public supervise."

The inception research was followed by a second phase (August 1999-August 2000) of further research and data collection. The Environmental Protection and Resources Conservation Committee of the National People's Congress collected, translated, and published materials of foreign cleaner production policies and regulations. In addition, wide and intensive study and research was carried out in preparing the legislation, with emphasis on the current situation in China, on the existing main problems in cleaner production stimulation and implementation, on the main constraints in cleaner production research and promotion, on the efficiency of cleaner production policies in China and abroad, and on feasible administrative systems applied in China and abroad. Four seminars were organized, attended by administrative departments of all sectors, research organizations, scientific and technological agencies, experts on environmental protection and resource conservation and legal experts. In addition various study tours were organized.[8]

Based on these investigations and meetings, the framework and outline of the Cleaner Production Promotion Law was developed. This rather thorough preparation solved or closed a number of key problems and debates in the drafting of the law:

- the main problems were defined on which the cleaner production legislation need to focus;
- the scope of cleaner production law was defined;
- the relation between the existing laws and the new cleaner production regulations were clarified;
- the applicable cleaner production encouragement policies were defined;
- the responsibilities of different administrative departments in cleaner production promotion were concluded.

The third phase consisted of drafting processes and discussions and started with the formulation of a proposal version by SETC on the basis of the studies that had been done. This report was set out widely for commenting and opinions at various governmental agencies (such as SEPA, State Development and Planning Commission, Ministry of Science and Technology, Finance Ministry) and in seminars for experts. This resulted in the first draft of cleaner production law in May 2001, three years after the start of the process. Further research and discussions on especially the legal aspects of the law contributed to the development of a second and a third draft in June 2001. The third draft was submitted to the plenary session of the EPRCC and amended into the forth draft. This so-called open-soliciting draft was distributed widely for comments and ideas to 49 departments of the State Council,

---

and methods in the final law have been formulated by SETC. Quite a few people in the consultation process argued for SEPA being the leading organization in implementing the Cleaner Production promotion Law.

[8] In this phase the drafting committee of EPRCC conducted study tours to Taiyuan Municipality, to France and to Canada. In addition surveys were made on Beijing Yanshan Petrochemical Corporation, on cleaner production application in Beijing Municipality, and on experiences in Anhui, Jiangsu, Shandong and Yunnan provinces.

all provinces, autonomous regions, municipalities, universities, research institutes and well-known experts. The feedback received from these bodies and the parallel research on local cleaner production legislation formed the basis for improvements made by the committee.[9] The fifth draft was submitted for approval to the plenary session of EPRCC. In December 2001, the members of the EPRCC finished discussions and amending and the final draft came into being, to be submitted to the Standing Committee of NPC for adoption.

The last step consisted of the review by the NPC. According to the arrangement of the Standing Committee, the NPC asked opinions on all aspects and also discussed the final draft with the Legal Committee and Legal Affairs Committee. On 29 June 2002, the Cleaner Production Promotion Law was adopted by the Standing Committee in the 28th session of the 9th National People's Congress of China. For the law, that conference was only its second reading, which can be interpreted as a very fast last track compared to other environmental laws.

Several reasons contributed to this rather smooth adoption of the law by the NPC. First, the nature of the law made it possible. The Cleaner Production Promotion law is based on incentives rather than on punishments and obligations as will be detailed below. Instead of direct government interventions in (industrial) production processes, the law stresses the importance of guidance, incentives and support to implement cleaner production in enterprises. The emphasis on "win-win" approaches was in line with such a legal strategy. Second, after ten years of awareness raising, experimenting and propaganda the relevant state and private organizations started to realize and accept the importance of cleaner production. Some major policy studies on cleaner production (see above) gave significant inputs in the law-making process and structured the most important policy departments in a common line of thinking. Third, China's accession to the WTO in 2001 made cleaner production more important for Chinese industry. To become competitive on the global market international standards on products and production processes were increasingly perceived as of high relevance. And last but not least, cleaner production, and especially in the framing of this Chinese law, was not really touching any serious interests of sectors or governmental departments.

4. The Contents of the Cleaner Production Promotion Law

The final law consists of 42 articles divided over six chapters. In this section we will review the contents of the law, including some of the discussions that took place in the law formulation process.

---

[9] By August 15, 2001 the EPRCC had received 67 reactions to their fourth draft. Some reactions were only or mainly positive, such as those of the Ministry of Foreign Affairs, the Congress of Tibet Autonomous region, the Congresses of Hainan and of Jianxi Province. Many others had detailed suggestions or comments, of which several made it into the law. Even at this stage of the law-making process some comments were rather frank in questioning the law as a whole. The Ministry of Agriculture doubted the necessity of this law, while the Congress of Zejiang province thought the conditions not ripe for such a law. The Supreme Court indicated that the provisions could be included in existing environmental laws.

Definition and Scope of the Law

During the law-making process a fierce debate took place on the definition and use of the concept of cleaner production. While initially the UNEP definition (as quoted in note 3) was used, finally a much more practical and - according to some - feasible definition was decided upon, to be found back in Article 2 of the Law:

Cleaner production as used in this Law means the continuous application of measures for design improvement, utilization of clean energy and raw materials, the implementation of advanced processes, technologies and equipment, improvement of management and comprehensive utilization of resources to reduce pollution at source, enhance the rates of resource utilization efficiency, reduce or avoid pollution generation and discharge in the course of production, provision of services and product use, so as to decrease harm to the health of human beings and the environment.

One major change that came out of the round of open comments on the fourth draft was a change in the name of the law. While initially the law was indicated with the Cleaner Production law, following comments of SETC the name was changed in the Cleaner Production Promotion law. With that the emphasis of the law on support, encouragement and voluntary actions by enterprises, instead of regulatory actions by governments, became clear.[10] It was not by accident that SETC, and not SEPA, came with this turn to a more promotional, supportive and encouraging law, rather than a regulatory law. Other, less powerful, actors (such as the Congress of Jinan city) have commented on the too large emphasis on voluntary and encouraging measures, questioning the effectiveness of such measures.

After ten years of experimenting and implementation of cleaner production in China, the concept of cleaner production has broken the boundaries of conventional industry and is being considered usable in agriculture and service sectors. The scope of application used in the law relates back to the definition of UNEP, including not only industry but also agriculture and service sectors. Article 3 of Cleaner Production Promotion Law writes:

> Within the territory of the People's Republic of China, any units or individuals engaged in activities related to production or provision of services and their corresponding management agencies must organize and implement systems for cleaner production in accordance with the provisions therefore contained in this law.
>
> For implementation the current law only provides detailed regulations of cleaner production promotion and application for industrial sectors, while only some general guidelines for cleaner production application in agriculture (article 22) and service sectors (article 23) are given.[11] With this design the law not only satisfies current urgent needs of cleaner promotion in the field of industrial production, but also provides legal room for future cleaner production promotion in other fields.

---

[10] At least in Japan similar promotion laws in the field of environment are known, e.g. the Effective Resource Utilization promotion Law. In Denmark a new section (no. 10) was included in 1991 in the Environmental Protection Act, regulating the use of binding voluntary agreements.

[11] Especially the Ministry of Agriculture was not too happy with this new law. Compared to industrial production, the articles on agriculture (and also those on the service sector) are rather general, and need to be detailed further to have any impact in current production practices. But the law does provide the legal basis to do so.

However, no article relates to issues of individual consumption of citizens in daily life, leaving the consumption of products outside the scope of the law.[12] Article 6 does specify that:

> The nation is committed to encouraging social groups and citizens to participate in the dissemination of public awareness with respect to cleaner production through, education, popularization, implementation and supervision.

This article and the related article 15 on the education and dissemination of cleaner production were specifically introduced after suggestions during the round of opinion soliciting. Some departments claimed that cleaner production activities should include social organizations and the media, as they have a role in cleaner production propaganda and awareness raising at the grassroots level.

The Stipulations for the Government

Although the Cleaner Production Promotion Law departs from a strict regulatory style, state organizations have been given a central role in it. It is especially chapter 2 that clearly stipulates the detailed requirements and tasks of the government and the relevant administrative departments. These tasks include formulating pro-cleaner production policies, formulating cleaner production promotion plans, developing a regional recycling economy, providing technical information and technical assistance to the enterprises, organising technical studies and technical demonstrations, and organising cleaner production education and propaganda.[13] The aims of these stipulations are that state organizations and administrative departments at all levels take measures to provide support and services for the enterprises who voluntarily initiate cleaner production. With that the state should create a good external environment for enterprises to start introducing cleaner production. As cleaner production is not simply pollution control, it cannot be achieved by relying on the management and supervision of only one department, as is usually the case in most of the environmental laws. Because of this and the fact that administrative fragmentation has always been an obstacle of legal implementation in China, the Cleaner Production Promotion law stresses the need for intimate co-operation among the relevant departments. Various departments need to be involved actively in the implementation, by their guidance, encouragement, support and regulation towards producers and managers from different perspectives and in different phases of the production process. Besides the law is also rather clear in defining different tasks and responsibilities of the various administrative departments within the state, trying to prevent bureaucratic controversies.

But the law leaves no uncertainty regarding the leading department in Cleaner Production Promotion law implementation. Article 5 of the law stipulates:

---

[12] The Congress of Shanghai Municipality commented that articles relating to green consumerism and recycling economy should be added, but this suggestion did not make it into the final version of the law.

[13] As Shi (2003, pp. 74-75) notes the Law regularly uses 'should' instead of stronger terms like 'must' or 'have to.' This is typically a consequence of the fact that this law tries to combine moral ethics with legal obligations. Such a law has to be followed by implementation regulations to specify the various obligations.

> The state department for economic and trade subject to the authority of the State Council shall bear the responsibility for undertaking the work of organising and coordinating the promotion of cleaner production throughout the nation. The relevant administrative departments responsible for environmental protection, planning, science and technology, agriculture, construction, water conservation and quality and technical supervision under the State Council shall assume the responsibility for promoting cleaner production in accordance with and pursuant to their respective functions and responsibilities.

The State Economic and Trade Commission and economic and trade commissions at the county level and above should do so by:

- formulating plans for promoting cleaner production;
- organising and supporting the development of information systems and technical advisory service systems for promoting cleaner production;
- promulgating directories of cleaner production technologies, processes, equipment and products on a regular basis;
- formulating and publishing directories of obsolete technologies, processes, equipment and products to be phased-out within set timeframes;
- designating a list of products subject to a mandatory material labelling scheme;
- formulating a list and implementing regulations for a compulsory recycling of products and packaging materials;
- formulating cleaner production auditing methods, in cooperation with SEPA, and reviewing the cleaner production audit reports of polluting enterprises;
- establishing voluntary agreements with enterprises concerning additional resource conservation and pollution beyond compliance.

While the SETC was given some tasks in other environmental laws, also with respect to pollution prevention and recycling, the Cleaner Production Promotion law for the first time gives SETC such a leading role and responsibilities in the field of environmental policy.[14]

The State Environmental Protection Administration, and the environmental protection bureaus at the county level and above, also play a major role within the Cleaner Production Promotion law, although they are not the main coordinating agency. Their roles are stipulated as:

- organising and supporting the development of information systems and technical advisory service systems for promoting cleaner production;
- reinforcing the supervision on the implementation of cleaner production in enterprises, and making public (via the media) the list of enterprises, which cannot satisfy the pollutant emission standards or mass-based pollutant discharge limits;
- ensuring that environmental impact assessments incorporate alternatives of cleaner production and audit reports are made of polluting enterprises;

[14] For instance, in the Resources and Energy Saving Law, the Environmental Protection Law, the Air Pollution Prevention Law and the Solid Waste Pollution Prevention Law.

- establishing voluntary agreements with enterprises concerning additional resource conservation and pollution reduction beyond compliance.

Moreover, it is noteworthy to mention that the law legalizes the delinking of SEPA and EPBs on the one hand from the affiliated business in charge of among others environmental impact assessments, environmental technology and ISO 14001 implementation, and environmental equipment manufacturing. This not only reduces the direct intervention of EPBs in enterprises' internal environmental protection decisions, but also levels the playing field for all enterprises involved in environmental industry.

Next to these two main institutions a number of other governmental organizations are mentioned in the law. For the State Development Planning Commission, the responsibilities are more restricted, including assistance in the formulation of cleaner production promotion plans and supporting the development of information dissemination and technical service capacities. The Ministry of Science and Technology and science and technology bureaus at the county level and above have responsibilities regarding organising and supporting the development of information systems and technical advisory service systems for promoting cleaner production, regarding directing supportive research, development, demonstration and diffusion of cleaner production technology, as well as regarding research and development of environmentally sound products. The Ministry of Agriculture and agricultural bureaus at the county level and above, should take a leading role in organising and supporting the development of information systems and technical advisory service systems for promoting cleaner production in agriculture, and organising the formulation of cleaner production guidelines and technical manuals. State Administration of Quality, Technology Supervision and Quarantine and quality and technology supervision bureaus at the county level or above are given the task to formulate technical codes for labelling the material contents of large-scale mechanical and electric equipment, mobile transportation vehicles, and any other products designated by the SETC. They also play a major role in the inspection and enforcement of the compulsory material labelling scheme as stipulated in Article 21 of the law. And they inspect and enforce the illegal activities associated with production and purchase of toxic and dangerous construction and decorative materials. Taxation authorities are important in deduction or exemption of the value-added tax for products made out of waste or raw materials recovered from wastes (see article 35). The Ministry of Education and their equivalents at lower levels are requested to incorporate cleaner production technology and management into the curricula of higher education, vocational education and technical training systems.

Various government departments may approve, when needed, the establishment of various labelling schemes for energy-efficient, water saving and recycling products and work out respective qualification procedures and standards. Equally, the organization of publicity and training activities on cleaner production, in order to raise the awareness of government workers, industrial managers and the public on cleaner production is spread among various governmental departments.

The Stipulations for Producers

The third chapter outlines the requirements and tasks given to the producers. The requirements can be divided into three main categories: general regulations guiding the implementation of cleaner production, mandatory regulation and voluntary regulation.

The guidance regulations (see Box 1) entail no legal responsibilities nor specific implementation rules, but put forward the kind of processes and measures through which cleaner production can be enhanced in companies.

Voluntary regulations (see Box 2) aim to encourage enterprises to initiate cleaner production voluntarily, improve the images of the enterprises and their products, through which they can get awards and enjoy favourable policies.

Box 1. Examples of Guideline Regulations

- Environmental impact assessments for new construction, renovation and expansion projects should review cleaner production options. (Article 18)
- Enterprises should use toxic-free/harmless or less toxic/less harmful raw materials; adopt high efficiency, less polluting technologies and equipment; reuse or recycle wastes, wastewater, and waste heat; and adopt pollution prevention and control technology to meet environmental regulation. (Article 19)
- Design of products and packaging should take the life-cycle impacts on human health and the environment into account and give preference to less toxic, degradable and recyclable options. Enterprises should adopt rational packaging practices to reduce packaging wastes. (Article 20)
- Agricultural producers should scientifically use fertilizers, pesticides, films and feedstock additives to efficiently produce high-quality agro-products and to recycle agro-wastes. (Article 22)
- Service facilities should apply energy/water efficient equipment and consumables. (Article 23)
- The construction sector should apply green design, materials and equipment. (Article 24)
- Exploration and extraction of mineral resources should adopt environmentally sound technologies and processes. (Article 25)
- Enterprises should reuse/recycle residues and waste heat by themselves or transfer it to other enterprises and individuals that can recycle them. (Article 26)

Box 2. Examples of Voluntary Regulations

- Establishment of voluntary agreements to further conserve resources and reduce pollution beyond discharge threshold standards with economic and trade commissions or environmental protection bureaus. These governmental agencies shall publish the company accomplishments under the voluntary agreements. (Article 29)
- Participation in accredited environmental management system certifications. (Article 30)

Mandatory regulations are essential duties for the producers, which include compulsory reuse of some products and compulsory labelling of some products or packages. Some heavily polluting enterprises are required to go through environmental auditing and regularly report on pollution emissions (see Box 3). Guidance regulations take the largest percentage of the articles, while mandatory requirements take only limited space. Subsequently the draft highlights the characteristics of a "promotion law," and reduces the idea of "administrative enforcement." Obviously promotion is considered to be more effective in leading producers to apply cleaner production and 'regulate' their activities into more environmentally sound paths.

Box 3. Examples of Mandatory Regulation

- Large electric and mechanical equipment, mobile transport equipment and other products designated by SETC are subject to a mandatory material labelling scheme. (Article 21)
- Toxic and dangerous wastes are banned for use as fertilizer and land reclamation (Article 22)
- Toxic and hazardous construction and decoration materials are banned. (Article 24)
- Designated products and packaging are subject to a mandatory recycling scheme. SETC will formulate the compulsory recycling directory, as well as the methods for recycling (Article 27)
- Enterprises using toxic or hazardous materials are subject to mandatory cleaner production auditing. (Article 28)
- Certain polluting enterprises shall disclose information on pollution release. (Article 29)

5. Inducement and Legal Liabilities

The Cleaner Production Promotion law has two separate chapters on measures, distinguishing encouragement (chapter 4) from legal liabilities (chapter 5).

Encouragement Measures

Chapter four stipulates measures to encourage enterprises to install cleaner production into their daily routines, such as awards, subsidies, favourable loans, and tax reductions, and government procurement. The origin and considerations for these measures are related to the problems businesses have met in cleaner production implementation in China, such as financial constraints, difficulties of starting cleaner production projects and lack of motivation. Five articles in chapter 4 specify the following measures encouraging cleaner production implementation:

- The state will establish awards for cleaner production.[15] The People's governments at various levels reward the institutions and individuals that accomplish substantial progress in cleaner production (Article 32);
- Cleaner Production research, demonstration projects and training programs, key projects of cleaner production technology upgrading and projects falling into voluntary agreements on pollution reduction are eligible for financial support through technology advancement funds that are allocated by the respective governments (Article 33).[16]
- An appropriate portion of the Small and Medium-sized Enterprise Development Fund (established under the Law on Small and Medium-sized Enterprises) should be allocated to support implementation of cleaner production by small and medium-sized enterprises (Article 34); SMEs who install cleaner production or conduct a cleaner production audit may apply for low-interest or noninterest loans from this foundation administrative agency.[17]
- Enterprises that make products, or recover raw materials, out of wastes are eligible for reduction or exemption of value-added tax in accordance with relevant government regulation (Article 35).
- The cost for enterprise cleaner production training and auditing can be accounted as production costs (Article 36).

Legal Liabilities

Chapter five stipulates the legal liabilities for noncompliance. Due the promotion character of this law it should not be too surprising that there are few articles on legal liabilities. But the fact that there are indeed enforcement articles distinguishes this Cleaner Production Promotion law from any promotional policy that is standard in western countries. There are five articles of noncompliance liabilities in the Cleaner Production Promotion law:

- Quality and technology supervision bureaus at the county level or above may impose a fine of up to RMB 50,000 (±Euro 6,250) on any enterprise that refuses to fully fulfill the obligation associated with mandatory material labelling (following article 21) within a preset deadline (article 37).
- Any entity that produces and sells construction and decoration materials containing toxic or hazardous substances exceeding the relevant national standards will be charged with administrative, civil or criminal penalties in accordance with the Law on Production quality and other applicable civil and criminal laws (article 38).

[15] The idea of a special Cleaner Production Prize, as suggested by some of the Standing Committee members, did not make it into the final draft.

[16] Initially, the idea was to have special funds established by all governments above county level to stimulate cleaner production. Several departments suggested that financial support could be paid out of the existing special funds, and they managed in that way to limit additional claims on their budget.

[17] The law especially made financial stipulations for SMEs, who meet special difficulties in getting commercial loans, influencing strongly their cleaner production implementation. This article was strongly discussed in the law-making process, especially from where the money should come from and who could decide on it. It was the Law Affair's Committee who succeeded in getting the article passed in this form.

- Economic and trade commissions at the county level or above may impose a fine of up to RMB 100,000 (±Euro 12,500) on any entity that refuses to fully fulfil the compulsory product and packaging material recycling obligations within a preset deadline (article 39).
- Environmental protection bureaus at the county level or above may impose a fine of up to RMB 100,000 (±Euro 12,500) on any enterprise that refuses to fully fulfil mandatory cleaner production auditing obligations within a preset deadline (article 40).
- Environmental protection Bureaus at the county level or above may impose a fine of up to RMB 100,000 (±Euro 12,500) on any entity that fails to fulfill pollution release information disclosure obligations in accordance with relevant requirements (article 41).

The penalty fees are comparable to those used in other environmental laws in China. Usually the liability fees are within the range of 30,000 to 100,000 RMB for violating enterprises. A fine above RMB 50,000 can be considered large and influential for most medium- and small-sized enterprise, and many penalties are well under RMB 50,000 if the violations are not very serious. For large scale enterprises (and especially those with foreign capital) fines up to RMB 100,000 do not have any serious impact on enterprise behaviour. A few interviews with leading managers of large enterprises in the chemical and petrochemical sector indeed confirm this (Liu, 2003). During the legislative process, arguments were raised to increase the levies and penalty fees to make the economic incentives stronger, without much effect (Du et al., 2001).

6. Implementation Assessment: From Cleaner Production to Circular Economy

As the Cleaner Production Promotion Law came into force in January 2003 it should be possible to give some first indications of its implementation trajectory and its impact in the practices of cleaner production. For several reasons, only some initial qualitative indications can be given, while an overall assessment proves rather difficult.

First, until now the law has not been subject to any serious assessment or evaluation. The Environmental Protection and Resources Conservation Committee of the National People's Congress of China has not included the law on its priority list of enforcement supervision. And also from the scientific community, little attention has been paid to assessing its effects and effectiveness, since it entered into force. There is evidence that cleaner production has increasingly made it into the industrial sectors. Cleaner production standards have been certified and implemented, for instance for tanning and oil refining, while for another ten industrial sectors CP standards are in the process of certification. For 17 other sector standards are being developed (Zhang, 2004).[18] At a local level there is increased evidence of systematic cleaner production auditing by enterprises, although actual implementation of the

[18] The National Development and Reform Commission, in which SETC and the State Development Planning Commission have been merged recently, has taken the lead in developing these standards, new methods for auditing.

measures listed falls behind. In addition, several initiatives by SETC and NDRC (the successor of SETC after a new government reorganization) and its local equivalents in several provinces and municipalities are further detailing the law and give it more regulatory possibilities.[19] But on the whole, it proves to be rather difficult to assess, for instance, the use of the various financial incentives or to get a full picture of law implementation and environmental effectiveness.

Second, the fact that the law closely interrelates with many other environmental laws in its implementation, all part of a much broader process of greening China's industrial and economic structure (see Appendix 1), make that the specific contribution of the Cleaner Production Promotion Law to actual transformations in production practices and arrangements is difficult to assess. Which parts of the industrial environmental improvements can really be related to the Cleaner Production Promotion Law and not to other regulatory initiatives?

An *ex ante* evaluation of the effectiveness and the impact of the law would draw strongly on the fact that this law is basically a promotional law, similar to, for instance, the Resources and Energy Saving Law. The voluntary nature of many articles, the fact that many responsibilities are with SETC, the limited financial arrangements, and the not too stringent obligations might make one rather skeptical regarding the direct contribution of this law to the greening of China's economy. There is the potential danger of what Ho and Vermeer termed the "greenwash" rather than the "greening" of state and society (Ho and Vermeer, 2006).

Many other Chinese environmental laws are more demanding, and the questioning of the necessity of this new law – as pointed out above – may seem rational. But the mere existence of this law sends out a strong signal to the business community, to the international community and to the different state and party organs. In line with this, it seems that the law is especially a codification of existing practices rather than working toward modifications and new practices. The wide-ranging experiments with cleaner production before the idea of the law was born; the endless cleaner production reporting, dissemination and training; and the wide trajectory of making the law, caused perhaps the largest effects within research, consultancies, international aid programs, and companies. Assessing the impact of the law should than include the period before the law came into force.

Finally, such an assessment is especially further complicated because in contemporary China environmental attention of authorities, academics and environmental experts has largely shifted from cleaner production to what is now labelled circular economy. Following the examples of the Japanese and German Recycling Economy laws, and intellectually inspired by the notion of Industrial Ecology (Shi et al., 2003), Chinese academics and leaders have strongly put the idea of circular economy on the research and policy agendas. Cleaner production has been almost completely integrated into ideas, research programs and experimental practices of circular economy.[20] China's 11th Five Year plan on the national economy has identified circular economy as one of the key elements, and the National

---

[19] H.H. Oliver and L. Ortolano (2004), *Promoting Cleaner Production at Industries in Jiangsu, China*, Presentation at the conference Energy and Environmental Awareness in China, Rice University, June 30, 2004. The provinces of Zhejiang and Liaoning have actively promoted clearner production in their jurisdictions.

[20] See for instance the recent overview of China's Circular Economy Initiative at: http:///www.indigodev.com/ circular1.html. And the work of the Task Force on Strategy and Mechanism Study for Promoting of Circular Economy and Cleaner Production in China, 2003 (a task force of the China Council for International Cooperation on Industrial Development): http://eng/cciced.org/cn/.

People's Congress is preparing the making of a circular economy law. And although the NDRC is warning that with the emergence of circular economy, the practice of cleaner production should not be put on the back seat, one cannot help to get the impression that this is happening.[21]

Overall, the general idea that emerges from the various interviews we had with state officials and experts on cleaner production is one of decreasing interest in and attention to the law now that it has entered into force. Few were really interested in finding out the actual results of the law. This does not mean that the Cleaner Production Promotion Law has been ineffective until now; future systematic evaluations on especially the local level have to investigate that. But it does point us to the fact that current developments in China, also within the field of environmental policy and management, go very fast; even to the point that serious interests in monitoring implementation and enforcement might be jeopardized as new agendas emerge before the old ones have stabilized.

CONCLUSION

With its high level of industrial and economic development, China is confronted with serious environmental problems and cleaner production was introduced into China as part of an answer to that. After more than ten years of propaganda, experimenting and project implementation, cleaner production gradually became accepted in government, party and industrial circles. This acceptance contributed to and enabled a rather smooth law-making process to institutionalise cleaner production in the policy and legal system. But the smooth law-making process was equally facilitated by some specific substantial characteristics of the law: a promotional law, broadly formulated in a flexible manner with little new and challenging legal and administrative obligations for environmental polluters. This substantive nature of the cleaner production promotion law reduced interests of the sectoral ministries to block the law. While the Cleaner Production Promotion Law is unique in that there is no other country that has a cleaner production law, the nature of the law very much resembles the stimulating cleaner production policies that we find in most of the industrializing and industrialized countries around the world. In that sense, China is no exception to the common belief that cleaner production can only be achieved within industries, and governments just have to set an enabling environment to promote, stimulate and facilitate the switch from curative and end-of-pipe practices to more preventive cleaner production. From an international perspective, the form rather than the substance of Chinese cleaner production policy draws attention.

[21] National Development and Reform Commission (2005), *The Circular (Recycling) Economy in China*, Beijing: NDRC.

Appendix 1
Cleaner Production in Other (Environmental) Laws in China

1. Some stipulations encourage scientific and technological innovation in the general articles of various environmental and resources laws.
2. The Resources and Energy Saving Law stipulates the broad responsibilities of the government and the energy saving requirements for enterprises. Specifically, the relevant regulations are:

    - Government organizes and implements big energy saving scientific research projects and pilot projects to encourage the adoption of advanced energy saving designs, technologies, equipment and raw materials;
    - Those new construction investment projects, who have not reached the rational energy consumption standards or do not comply with the energy saving principles, will not be ratified;
    - Energy using units should launch education, propaganda and training on energy saving; apply the energy saving standards; and eliminate the old high energy intensive products, technologies and materials;
    - The legal liabilities are: governments above county level can command to stop high energy intensive industrial projects; they can install a timeframe for improvement or even stop the enterprises that consume energy exceeding the standards; and they can charge fees on enterprises which produce obsolete products or equipment forbidden by state.

3. In Environmental Protection Law of PR China, the cleaner production relevant stipulations are:

    - Pollution generation units should include environmental protection in their work agenda, adopt effective measures preventing the generation of waste gas, waste water, waste slug, suspended particles, odor, radioactive materials, noise, etc. that will cause environmental pollution and damages;
    - Newly constructed enterprises or industries should adopt the equipment and designs of high energy utilization rate and low pollutants generation, and adopt economic and rational waste comprehensive utilization technologies;
    - The pollution prevention and control equipment of construction projects must be designed, built and put in use simultaneously with the main project (three simultaneous).

4. Scientific and Technological Innovation Application Law has stipulations on promotion of application of scientific and technological innovation on resources and energy saving, and on environmental protection.
5. Air Pollution Prevention and Control Law has the following cleaner production stipulations:

- Governments of all levels should award the units that fulfill remarkable achievement in air pollution prevention and control, and protection and improvement of air environment;
- The state encourages and supports the scientific research on air pollution prevention and control, and spreads the advanced and suitable air pollution prevention and control technologies;
- Enterprises should preferably adopt cleaner production designs of high energy utilization rate, low pollution discharge to reduce generation of air pollutants. For some obsolete designs and equipment the state will make timeframes of eliminating them. The State Council will publish the directories of the forbidden designs and equipments that will cause serious air pollution;
- The State Council and governments of all levels should adopt measures to improve the urban resources structure and spread widely the production and application of clean resources;
- The state adopts the economic and technological policies and measures that are in favor of clean utilization of coal; encourages and supports the development and spreading of clean coal technologies;
- The flaming gas generated from industrial production should be recovered and utilized.

6. Solid Waste Pollution Prevention and Control Law have regulated the solid waste prevention, using of toxic and harmful materials and pollutant discharge limitations. The articles are the followings:

    - The state encourages and supports resource comprehensive utilization, and recovery and rationally utilization of solid wastes;
    - Packaging of products should be easy to recover and treat;
    - The state should organize and spread the advanced production designs and equipment preventing industrial solid waste pollution;
    - The state publishes the directories of obsolete designs and equipment that generates seriously polluting industrial solid wastes.

7. Water Pollution Prevention and Control Law has three stipulations on cleaner production:

    - The State Council and governments of all levels should rectify the enterprises causing heavy water pollution and force them to technological innovation: adopt comprehensive prevention and control measures and reduce the discharge amount of waste water and wastes;
    - Enterprises should adopt clean production designs with high resources utilization efficiency, low pollutant discharge, and strengthened management;
    - The state forbids establishing SMEs of paper, textile, paints, leather, petrochemical and pesticide industries without water pollution prevention and control measures.

8. Marine Environmental Protection Law also has two stipulations on cleaner production:

    - The state should strengthen the scientific and technological research on preventing and controlling marine environmental protection; it should eliminate obsolete designs and equipment;
    - It is forbidden to establish industrial projects of paper, chemical, textile, leather, brewery, and other projects that may cause heavy pollution to the marine environment.

REFERENCES

ADB (2001), *Report of B-4 Project*, Manila: Asian Development Bank, B-4 expert team.

Du, X., F. Hao, H. Zhang and Y. Wang (2001), "The proposal to the policy and law system for Cleaner Production in China." Paper presented at the International Conference on Cleaner production, September 2001, Beijing, China (also at: http://www.chinacp.com., October 2003).

Ho, P. and E. Vermeer (2006) (eds), *China's Limits to Growth: Prospects for Greening State and Society*, Oxford: Blackwell Publishers.

Liu, Y (2003), *Cleaner Production Promotion Law of China*, Wageningen: Wageningen University (MSc thesis).

Shi, H. (2003), "Cleaner Production in China." In: A.P.J. Mol and J.C.L. van Buuren (eds), *Greening Industrialization in Asian Transitional Economies: China and Vietnam*, Lanham: Lexington, pp.61-82.

Shi, H., Y. Moriguichi and J. Yang (2003), 'Industrial Ecology in China, Part 1: Research,' *Journal of Industrial Ecology* 6 (3-4): 7-11.

Zhang, H. (2004), *Cleaner Production (CP) in China*, CTI workshop presentation, Beijing: SEPA.

Zhang, T. (2003), Personal communication, Beijing: Tsinghua University.

Zhang, T. (2005), Personal communication, Beijing: Tsinghua University.

Zhang, T. and J. Chen (2001), "Promoting Cleaner Production in China." Paper presented at International Conference on Cleaner production, September 2001, Beijing, China (also at: http://www.chinacp.com).

In: Leapfrogging Development in Emerging Asia
Editor: Peter Ho, pp. 39-61
ISBN: 978-1-60456-410-5

Chapter 3

ENVIRONMENTAL LEAPFROGGING THROUGH VOLUNTARY MEASURES? ISO 14001 IN CHINA

***Katherine Kao Cushing*[1]*, Heather McGray*[2] *and Hongyan Lu*[3]**
[1]Environmental Studies Department at San Jose State University.
[2]ECOLOGIA, Vermont, USA
[3]Institute of Applied Material Flow Management, Trier University

In the debates on environmental leapfrogging and greening, ensuring the cooperation of industries through the implementation of voluntary measures, rather than imposing policies of "command-and-control" is deemed essential (Ho and Vermeer, 2006). This chapter explores the scope and possibilities for such measures by narrowing down on one of the largest emerging economies in Asia: China. Since the 1990s the government of the People's Republic of China has been actively promoting the use of voluntary methods for improving the environmental performance of enterprises. Environmental Management Systems (EMSs)—procedure-based tools for systematically identifying, monitoring, and strategically managing all of an enterprise's environmental impacts—are viewed as an important new way to promote pollution control and cleaner production in China. It hass claimed to align China's environmental protection efforts with international standards, and to provide Chinese enterprises with environmental credibility in the international marketplace (China People's Daily 2000). Since 1996, the number of enterprises adopting EMSs has risen exponentially. ISO 14001, an EMS developed by the International Organization for Standardization (ISO) in Switzerland, is the most well known international EMS standard, and the one that has been declared the "official" EMS of China.

In 2002 China ranked fifth in the total number of ISO 14001 certifications, with 2,803 firms certified to the standard (ISO 2002), not including certified enterprises in Hong Kong (208) and Macau (70), which had about the same number of certified enterprises as the U.S., the U.K., and Germany. Regionally, China ranked a distant second in the number of certified enterprises, behind Japan's 10,620 certifications.

Other Asian countries with a significant number of certifications included South Korea (1,065), Thailand (671), and Singapore (441). Compared to other world regions, Asia ranks second in terms of percent of total certifications, making up about 36% of all certified firms, compared to Europe with 47%. By June 2004, the number of certifications in China had increased by 134%, with 6,546 enterprises certified (CNAB 2004).

Clearly, the EMS activity in China is increasing rapidly and shows no signs of abating. However, very little is known about how much ISO 14001 certification does to quantitatively improve an organization's environmental performance. Is this voluntary management tool, which lacks any specific performance requirements, really a viable option for demonstrably improving environmental performance in a country that has trouble enforcing existing mandatory environmental regulations?

In response to this overarching question, this chapter investigates five areas:

1. What is the current implementation status of the ISO 14001 in China? How does this compare to adoption rates in other Asian countries?
2. What role does the Chinese government play in the certification of enterprises?
3. What factors influence the adoption of ISO 14001 by Chinese enterprises?
4. What effects have ISO 14001 had on the enterprise performance? What have been the main benefits and difficulties?
5. How do larger institutional factors, such as China's existing environmental regulatory framework, influence ISO 14001 implementation?

1. ISO 14001 Background

EMSs: Definition and Potential Benefits

An EMS is a set of organization-specific policies, procedures, and guidelines that govern the strategic and day-to-day operations of a company's environmental activities. In the U.S., these activities have traditionally focused on the compliance with wastewater, air pollution, solid and hazardous waste, and worker's health and safety regulations. More recently, however, more and more businesses have used EMSs as proactive tools for addressing both regulated and non-regulated aspects of their business. For example, Hewlett Packard's local site operations use their EMSs to help them identify site specific environmental, health and safety issues, set goals and implement improvement plans, and implement preventative actions (HP 2004). Additionally, EMSs can be used to help organizations integrate the environmental aspects of their business with product design, manufacturing, and marketing (e.g., through improved raw material use, process efficiency, and the adoption of cleaner production activities) (Lamprecht 1997; Melnyk et al. 1999). EMSs are also tools for identifying cost saving opportunities and for mitigating environmental risk that may be of concern to management or investors (Morrison et al. 2000).

The ISO 14001 EMS standard, the world's most common EMS framework, is comprised of the following iterative parts: policies, planning, management programs, internal auditing, and operations control. Table 1 below briefly outlines the function of each of these steps.

Once a company has established its own EMS, it may elect to hire a third party, or registrar, to conduct an independent assessment. When a registrar organization is hired by a firm to conduct an audit, it sends out a team of EMS auditors, a group of individuals professionally trained in the details of the ISO 14001 implementation. If the registrar finds that the company's EMS is in conformance with the ISO standard, the company receives "certification" to the ISO 14001 standard.

Table 1. The EMS Process

Phases	Description
Environmental Policy	Top-level organizational management establishes a policy that must contain commitments to at least the following items: prevention of pollution, continuous improvement, regulatory compliance, framework for setting and reviewing environmental objectives and targets. The policy must be made available to employees and the public.
Planning	The organization identifies all of its significant environmental "aspects," for example SO_2 emissions or COD discharges. Once identified, these aspects form the basis for target setting, monitoring, and management controls. In this phase, the organization assigns responsibility for setting and meeting objectives to specific departments and people within the company.
Implementation and Operation	In this phase, the organization further specifies the responsibilities of individual departments and personnel for monitoring and managing specific environmental aspects. This phase often involves training employees and documenting or creating documentation for procedures related to the management of the company's specific environmental aspects.
Checking and Corrective Action	On a regular basis, the organization monitors and measures its progress with respect to meeting it own environmental goals, which includes regulatory compliance. The company must also have or develop a procedure for identifying non-conformance and plans for corrective and preventative action. This stage involves conducting periodic, comprehensive internal audits of the EMS and keeping detailed records.
Management Review	EMS audit reports are communicated to top-level management, who determine whether or not the company's EMS is performing adequately.

Source: ISO 1996; Morrison et al. 2000.

Good faith implementation of an EMS can have dramatic positive impacts on an individual organization's environmental performance. U.S. companies that have implemented ISO 14001 have demonstrated improved relations with environmental regulators, energy and raw material cost savings, reduced waste products, and reduced insurance costs (Melnyk et al. 1999; Morrison et al. 2000). In a survey of 1,401 American manufacturers, researchers found that benefits to companies that were in the process of or were already 14001-certified were real and significant, but that there was no comparable data that evaluated the ratio between dollar outlays for implementing a company ISO 14001 program and direct benefits. In addition, many of the benefits companies reported were qualitative in nature (e.g., making the company think more proactively about its environmental issues), making it difficult, if not impossible, to place a dollar value on them (Melnyk et al. 1999). A major limitation of EMSs like ISO 14001 is that since certified organizations are complying only with a *management system*, as opposed to a set of *quantitative performance standards*, there is no guarantee 1) that a certified facility truly demonstrates good environmental performance or 2) that the facility's performance will actually improve over time (Morrison et al. 2000).

2. ADOPTION OF ISO 14001 AND THE CHINESE GOVERNMENT'S ROLE

In response to the first two research questions, this section of our chapter discusses the history and current status of ISO 14001 adoption in Chinese enterprises and compares these figures to certification rates in other industrializing Asian countries and globally as well. It also outlines the role of government agencies, such as the State Environmental Protection Agency (SEPA) and the China National Accreditation Board for Certifiers (CNAB).

Adoption Rates

Figure 1 below shows the number of enterprises with ISO 14001 certification to ISO 14001 standard from 1997 to 2002. As of December 2004, China had 6,546 enterprises that had obtained official third party certification of their EMSs to the ISO 14001 standard (CNAB 2004). This number is almost 30 times the number of enterprises certified in 1999. Between 1997 and 2000, the number of annual certifications grew on average 170% (ISO 2002).

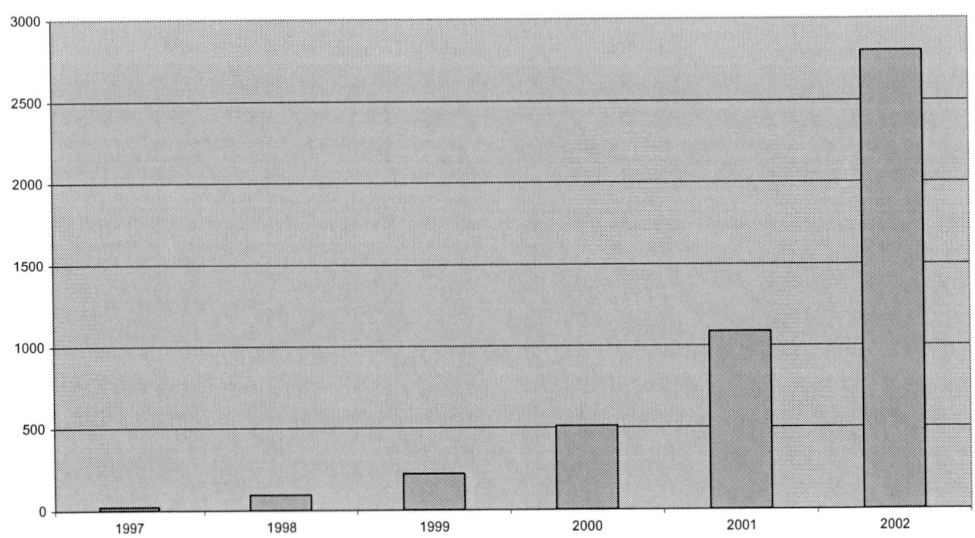

Source: ISO 2002.

Figure 1. ISO Certifications in China.

This pattern of exponential growth, albeit less extreme, is also occurring in other countries in Asia, such as Singapore, Thailand, and South Korea. For example, in both

Singapore and South Korea, the number of certifications increased over six times from 1997 to 2002; in Thailand, the increase was 11-fold (ISO 2002).

Figure 2 shows the number of ISO 14001 certificated enterprises by province or city. Not surprisingly, the locales with the most certifications are Guangdong with 215, Beijing with 137, and Jiangsu with 134. All of these areas are on the east coast, which corresponds with the concentration of industrial activity in China (CRCEA 2002).

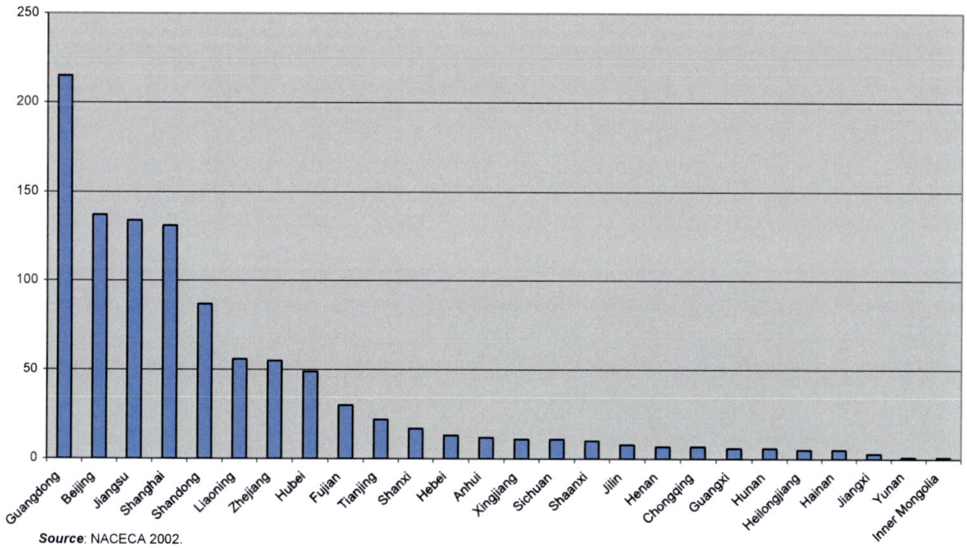

Figure 2. ISO 14001 Certifications by Location.

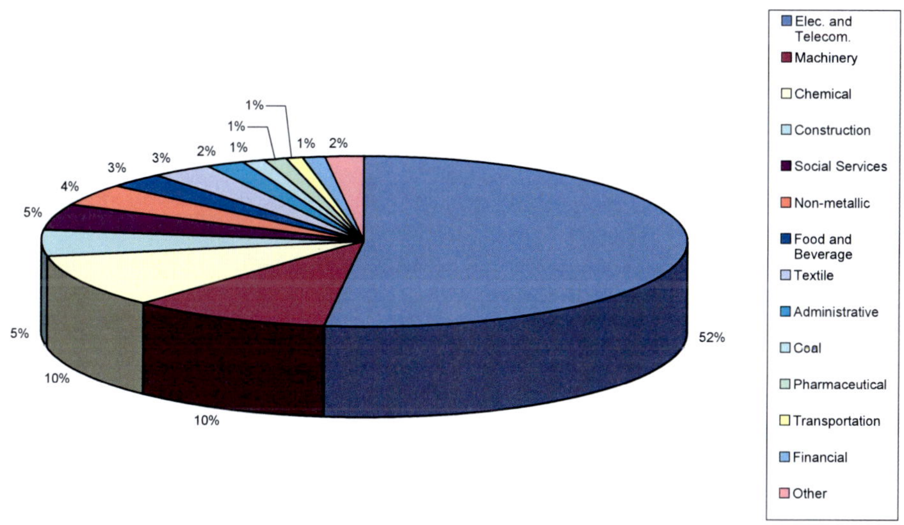

Source: NACECA 2002.

Figure 3. ISO 14001 Certifications by Industry.

However, interest in ISO 14001 certification has penetrated into even the most remote and nonindustrialized provinces. This is evidenced by the handful of facilities that have received certification in provinces such as Guangxi and Heilongjiang, with six and five certified facilities, respectively (CRCEA 2002).

Figure 3 provides a breakdown of ISO 14001 certifications by industrial sector as of July 2001. This chart indicates that one group—electronics and telecommunications—accounts for the vast majority of certifications (52%). Combined with certifications in the machinery (10%) and chemical (10%) industries, enterprises in these three manufacturing sectors make up over 70% of all certifications (NACECA 2001). Based on a survey of 108 certified Chinese enterprises Zeng and Wang (2002) characterize these adopters as mostly export-oriented. Additionally, the authors group enterprise ownership of certified facilities into five categories: joint venture (59%), state-owned (20%), share-holding (13%), private (4%), and other (4%).

Government Role and Institutional Setting

Following ISO's formal issuance of the ISO 14000 series of standards in 1996, the Chinese national government established an interministerial organization in 1997—the China Steering Committee of EMS Accreditation. The purpose of this organization was to oversee the implementation of ISO 14000 standards in China. The Steering Committee operates under the auspices of the State Environmental Protection Administration. Until 2003, the Steering Committee had two working groups that carried out the day-to-day work associated with getting facilities certified to the ISO 14001 standard, the China Accreditation Committee for EMS Certification Bodies (CACECB) and the China Registration Committee for Environmental Auditors (CRCEA). The CACECB is responsible for the approval and certification of registrars, while the CRCEA oversees the training, approval, and registration of ISO 14001 auditors (UNEP-IETC 2003).

Thirty-three organizations within China are accredited as registrars (NACECA 2001). Table 2. below lists several examples of certifying organizations and shows that they are found in a variety of organizational settings, ranging from industrial ministries to SEPA to branches of local environmental protection bureaus (NACECA 2001). Registrars can be loosely divided into two categories: those with an environmental background and those with a management background. Environment-oriented registrars tend to be located within environmental protection bureaus or governmental research institutes, where ISO 14000 certification is part of a suite of activities including research, environmental impact assessments and regulatory enforcement. Management-oriented registrars may be housed within a municipal economic development bureau or a provincial standardization institute, but are funded primarily through income from certification activities and may have the culture and climate of private companies. For them, ISO 14000 is part of a larger set of management certifications including ISO 9000 (quality management), OHSAS 18000 (occupational health and safety), and other national and international standards. These certification centers enable enterprises to undergo a single integrated certification process that results in multiple certificates. Such multiple certifications, especially dual ISO 9000/14000 certification, represent a growing share of ISO 14000 certifications in China (Zhang 2002).

Table 2. Examples of ISO 14001 Registrars in China

Name	Key Characteristics
Shenzhen Quality Certification Centre www.sqcc.com.cn	Shenzhen-based certification company with eight branches, including Zhuhai, Chengdu, Hangzhou, and Anshan, Liaoning. Primarily conducts ISO 14001, ISO 9000, and OHSAS18000 certifications, including integrated certificates. Characterizes itself as a "professional nonprofit organization."
Shenzhen EMS Certification Center www.szemscc.com.cn	Division of the Shenzhen City EPB. Serves exclusively as an environmental certifier, including a new green labeling project. Collaborates on joint 9000/14000 certification with auditors from other organizations.
CEPREI Certification Body www.ceprei.org	Located in Guangzhou. Conducts certification to many environmental and nonenvironmental standards. Has ISO 14001 accreditation under the US Registrar Accreditation Board, in addition to Chinese accreditation.
South China Certification Center of Environmental Management Systems	Small governmental certification center in Guangzhou. Housed at the South China Institute for Environmental Sciences, a regional research center of SEPA that does a variety of consulting, training, research work.

Many organizations that are not part of the Chinese government are also involved in the implementation of ISO 14001 in China. These include foreign registrars, such as BVQI and Excel, multilateral environmental organizations, such as United Nations Environmental Programme, and the national-level environmental organizations or aid organizations of individual countries, such as Canada, Germany and the U.S. multilateral organizations and national level environmental organizations are often key players in ISO 14001 demonstration projects, providing funding and technical expertise.

3. Factors Influencing Adoption

In investigating our third research question, "What factors influence adoption in Chinese enterprises?" we identified four main drivers: 1) international trade and transnational corporate policy, 2) government environmental improvement initiatives, 3) the potential for improved environmental performance and cost savings, and 4) top company management. In the following discussion, we discuss each of these factors in turn.

International Trade

Chinese enterprises want to compete in the international marketplace, both as independent providers of goods and suppliers to foreign companies. Certification, especially if it is performed by an international environmental consulting company, helps lend legitimacy to Chinese firms that want to be considered at the same quality level as manufacturers in developed countries. Moreover, a growing number of transnational companies, such as Ford Motor Company, DaimlerChrysler, and many Japanese electronics firms, are requiring that their major suppliers become ISO 14001-certified (SustainableBusiness.com 1999). Also, more and more government procurement policies, especially in Europe, give preference to ISO 14001-certified companies. More generally, the Chinese national government's support for ISO 14001 supports the argument that Chinese businesses can operate according to a global "rule of law," a trade issue in which China has been widely criticized (Spigelman 2002). Other reasons cited by Chinese enterprises and organizations as reasons for obtaining ISO 14001 certification that relate to international trade include attracting foreign investment (e.g., foreign companies establishing subsidiaries in their areas), conformance with new EU directives on reducing the use of hazardous substances in electronic products, and promoting the "green image" of a city, town or special economic zone (Kodak 1999; CCCPCP 2001; Geng and Cote 2003).

Transnational Corporate Policy

Hundreds of the world's largest transnational corporations (TNCs) have manufacturing operations in China. Many of these companies, such as Kodak, Ricoh, and Hewlett-Packard have their own set of corporate environmental policies, and they expect all their manufacturing facilities, regardless of location, to adhere to them (Kodak 1999; Ricoh 2002; Hewlett-Packard 2004). The inclusion of ISO 14001 in these policies is reflected in the high percentage of Chinese ISO 14001 certifications found in joint-venture enterprises. For example, Kodak (1999) states that one of its goals for all of its major manufacturing sites in China was to achieve ISO 14001 certification by 2001. The company has manufacturing facilities in Shanghai and Xiamen. Cushing (1998) also found that the European corporate headquarters of a transnational pharmaceutical firm exerted a strong degree of influence on the implementation of environmental programs in its Shanghai subsidiary facility, particularly with regards to encouraging the enterprise to comply with local wastewater treatment regulations.

Government Environmental Improvement Initiatives and Pilot Projects

Chinese government agencies, from the national to the local level, play a vital role in ISO 14001 certification, ranging from creating programs to attract early adopters to disseminating information about certification benefits to enterprises considering certification. A 2003 UNEP report identifies national-level governmental policies (e.g., SEPA pilot programs) as being a "particularly important factor for the adoption of ISO 14001 certification for Chinese enterprises and administrative areas" (UNEP-IETC 2003). For example, China's Tenth Five-

Year Plan emphasizes the need to build the nation's environmental management, consulting, and certification capacity, which has strengthened efforts by SEPA to promote ISO 14001 certification (Zhao 2002). Since the mid-1990s, SEPA has been actively encouraging enterprises and administrative areas to become certified and has promoted certification through a variety of training and demonstration projects. One of the main reasons for certification cited by a Copyer (a company that manufactures paper copiers) manufacturing subsidiary in Shenzhen was that the city was designated a "model city" by SEPA in 1997 (Copyer 2002). Similarly, project managers involved in Suzhou Industrial Park's ISO 14001 certification cited SEPA's "First National Demonstration District" program as an impetus behind their decision.

In addition to national level programs and projects, previous research (e.g., Sinkule and Ortolano 1995; Warren 1996; Cushing 1998; Ma and Ortolano 2000) demonstrates that *local* environmental improvement initiatives often play a major role in enterprise environmental management, and that they can be a critical factor in distinguishing between poor and good enterprise environmental performance. For example, one of the main reasons cited by the Dalian Economic and Technological Development Zone for obtaining certification was Dalian's "Blue Sky, Green Sea" Project, a local environmental cleanup initiative with the goal of protecting the local natural environment, enhancing local residents' living conditions and increasing the city's international economic competitiveness. Park officials in Dalian strongly believed that adopting the ISO 14001 standard would help them achieve this program's goals (Geng and Cote 2003).

A growing number of Chinese local environmental improvement initiatives focus explicitly on ISO 14001 certification, making it, in some cases, an end in itself, rather than a mechanism for achieving the goals of a particular initiative or program. The city of Shanghai has created an ISO 14001 incentive program that awards RMB 20,000 (approximately $2,400 US) to certified companies (Shenzhen EMS auditor 2002). In Beijing, SEPA has piloted district (*qu*)-level ISO 14001 certifications, where an EMS serves as a focal point for integrated district environmental planning (Teng 2003).

Potential for Regulatory, Economic, and Environmental Benefit

Within Chinese industry, there is a growing recognition that addressing environmental concerns can significantly impact operations, competitiveness, and profitability. Historically, noncompliance with environmental regulations at the enterprise level has often been pervasive (Ma and Ortolano 2000), and the existing legal system for the enforcement of Chinese environmental regulations has recently been characterized by enterprise management as "not strong" (Zeng and Wang 2002). Nevertheless, Chinese environmental officials have been known to take strong steps to close down facilities that were not properly treating their waste or to force enterprises with faulty equipment to redesign and retrofit existing pollution control equipment (Sinkule and Ortolano 1995; Economy 2004). EMSs like ISO 14001 provide a management tool to help enterprises identify and develop plans for dealing with environmental regulatory violations, as well as integrate new environmental policies into their operations.

EMS adoption, when combined with cleaner production (CP), holds great potential for an enterprise to realize concurrent economic and environmental benefits. For example, in

Suzhou Industrial Park, one facility that conducted a CP audit in conjunction with its ISO 14001 certification activities was able to realize a savings of 15 million RMB (approximately $1.8 million US) and significantly reduce energy consumption (CCCPCP 2001). Even when undertaken in the absence of an underlying EMS, CP projects by Chinese enterprises, have demonstrated significant benefits. Warren's (1996) study of CP in the Chinese electroplating industry and Cushing, Wise, and Hawes-Davis' (1999) investigation of CP in the Chinese pharmaceutical industry provide evidence that enterprises can recognize, act upon, and reap substantial monetary and economic benefits from CP opportunities identified through the audit process. Major components of CP audits (e.g., analysis of the enterprise's main waste streams) map well into the Plan-Do-Check-Act cycle that forms the basis of all EMSs. Cushing and colleagues (1999) also found, however, that while enterprises will implement most of the low and no-cost CP options identified through an audit, they are much more cautious about making medium- and high-cost investments in CP-related projects. In addition, they posit that the economic costs and benefits analyses conducted on the higher cost CP projects did not provide decision-makers with enough evidence to justify the investment.

The potential for China to make significant economic and environmental gains from the combination of environmental management systems and CP projects was given a boost with the passing of the national Cleaner Production Promotion Law (National People's Congress 2002), which came into effect in 2003. The law creates a range of incentives for companies and economic officials at all levels to find cost-effective ways of preventing pollution. In addition, the Cleaner Production Promotion Law is unusual among Chinese environmental legislation in that the State Development Reform Commission was given primary responsibility for its implementation rather than the State Environmental Protection Agency. This makes it likely that in coming years, EMS and CP activities will be promoted not only by the relatively weak SEPA and EPBs, but also by economic development agencies that often have more clout.

Top Company Management

Top-level company management plays a critical role in the implementation of an EMS. Said one corporate environmental manager, "The attitudes toward EMS of company leadership at the highest level highly influence EMS adoption and implementation," (Suzhou enterprise environmental manager 2004). In an additional survey of over 100 certified Chinese enterprises, Zeng and Wang (2002) asked top enterprise management to rank a variety of factors influencing the implementation of EMS at their company. The top three factors were the environmental consciousness of the company's top leaders, the responsibility of the company to manage environmental issues, and "legal system and legal enforcement," which presumably refers to compliance with existing environmental regulations. Interestingly, the factor of "minimizing environmental nuisance materials" only ranked eighth out of 27 factors. The results of this survey indicate that the certification process is largely driven by top company management, but do not provide insight into how or whether ISO 14001 certification actually resulted in improved environmental performance.

4. Implementation Results

The most detailed information assessing the effectiveness of EMSs comes from a handful of case studies, with the two most detailed case studies focusing on ISO 14001 certification in industrial parks in Suzhou and Dalian, as opposed to individual industrial facilities. Nevertheless, these cases provide useful insight into the potential benefits of an EMS at the enterprise level.

Table 3 summarizes some of the case studies more significant results. Together they illustrate the wide range of possible ISO 14001 certification outcomes. Furthermore, they demonstrate that certification can indeed contribute to significant improvements in enterprise environmental performance, both quantitatively and qualitatively. Broadly, the benefits fall into the following categories: cost savings, improved regulatory compliance, reduced resource use, pollution reduction, new procedures, infrastructure investment, and diffusion.

Cost savings were explicitly stated in two of the case studies—Suzhou and Shandong cases—as being significant. The Suzhou case directly attributed this cost savings to decreased energy, water, and wastewater treatment demands; for the Shandong case it was reduced water and coal consumption. Along the same lines, a Guangdong beer manufacturer cited water use reduction by 14% per kL of product (Guangdong enterprise environmental manager 2002). Energy and water use reduction were likewise noted in the Dalian case. The Dalian case study authors also reported a significant amount of pollution generation for priority pollutants. More specifically, they note a 46% decrease in the concentration of total suspended particulates (TSP) and a 17% reduction in the concentration of SO_2 emissions. As examples of procedural improvements, the Suzhou Industrial Park created new programs for monitoring chemical use and a flood emergency response plan. Improved regulatory compliance was also reported as a benefit by preliminary findings of a SEPA survey of 15 certified enterprises (Chen 2002).

The aforementioned examples of certification benefits are not unexpected—these kinds of paybacks are to be expected and are often touted by certification promoters as reasons for going through the certification process. What is more unusual and perhaps even more significant in the case of China is that through certification, there are some examples of significant monetary investments being made to improve an enterprise's or administrative area's waste management infrastructure. A specific example of this is the Dalian Economic and Technological Development Zone constructing a hazardous waste landfill and increasing the capacity of its wastewater treatment facility.

Another positive outcome of at least one pilot project that has broader significance involves the influence that the certification of one area had on encouraging other enterprises to become certified. In the case of Dalian, the certification of the administrative area served as the catalyst for 14 other facilities within the park to initiate the certification process. The idea that the desire for certification can spread independent of a formally sanctioned and managed government program indicates that enterprise managers in the zone recognized the benefits of EMSs outside of certain high profile "pilot programs." If true, this assertion bodes well for the self-sustaining diffusion of ISO 14001.

The third column of Table 3 entitled Data Status summarizes our evaluation of the quantity and quality of available data. It shows that, while the reported benefits can be substantial, important pieces of data are missing that are needed for any generalizable conclusions to be made about the relationship between certification benefits and costs.

Table 3. Summary of ISO 14001 Case Study Results

Site/Case Description	Result Highlights	Data Status
Suzhou Industrial Park	• Instituted chemical inventory program • Created flood control procedures • Created emergency response plan for flooding • Some significant CP benefits (e.g., energy and water savings totaling more than 15 x 10^6 RMB annually for facility within Park) • Eleven new public gardens developed • Added of 35,000 m^2 green space • Now complies with or surpasses national standards for air, water, noise, COD, and BOD	• No direct and indirect implementation and certification costs reported • Limited before and after comparison • Data not comparable across enterprises
Dalian Economic and Technical Development Zone	• Reduction of TSP concentration by 46% • Reduction of SO_2 concentration by 17% • Reduction of COD by 13.2% • Reduction of NO_x by 9% • Increased energy efficiency (e.g., by switching fuel from coal to liquefied petroleum gas and by converting to district heating) • Increased percentage of wastewater treated before discharge • Increased use of recycled water (1.2 million tons/year) • Constructed hazardous waste landfill • Increased capacity of wastewater treatment facility • Impetus behind 14 facilities within park to obtain ISO 14001 certification	• No direct and indirect implementation and certification costs • Some before and after comparison • Data not comparable across enterprises
Blue Sky Power Company (pseudonym), Shandong	• Reduced water use by 2,600,000 to 3,900,000 tons per year • Lowered coal consumption to 50 kg per 10,000 kilowatt hour of electricity produced • Improved coal ash recycling provides income of tens of millions of RMB per year. • Reduced dust and noise pollution	• More detailed before and after comparisons • Some cost savings metrics present

Source: CCCPC 2001; Geng and Cote 2003; Matouq, 2000; Huang, 2002; Blue Sky Power Company, 2004.

The most detailed case study, which does contain some before and after comparisons, comes from the Suzhou Industrial Park case study. This case study demonstrated significant quantified decreases in air pollution emissions (e.g., SO_2) and wastewater pollutant discharges (e.g., COD). However, across, the board, no detailed data was available on variables such as pollution discharges and resource use before and after certification. Nor was there any significant information on direct and indirect implementation and certification costs.

Chinese enterprises' lack of information on the effect certification had on overall corporate performance is not unique. For comparison's sake, even in an information-rich developed country like the U.S., the extent to which enterprise managers reported that obtaining certification improved various aspects of overall performance was generally lukewarm. For example, in a comprehensive survey of over 1,500 certified U.S. corporations from a variety of industrial sectors, Melnyk and colleagues (1999, p. 36) asked corporate management a series of questions on the standard's effects. More specifically, respondents were asked to rank on a scale of 0 (strongly disagree) to 10 (strongly agree) if they thought the benefits of certification definitely outweighed costs (score of 4.18/10.0) or if certification helped enhance the reputation of their company (score of 4.81/10.0). On both of these aspects, the mean score was below the middle value of 5.

The cost of EMS implementation remains an important unknown variable. Consulting fees associated with establishing an EMS typically amount to RMB 100,000 (about $12,000 US) for Chinese consultancies (Teng 2003) and are often more than double that for foreign consulting companies (Qian and Li 2002). However, EMS consulting may represent a relatively small share of EMS implementation costs. Without data on implementation costs, an overall assessment of EMS effectiveness for these cases is infeasible.

Regarding challenges in EMS implementation, managers maintain that it is easier to obtain ISO 14001 if a company already has an EMS than if one must be started from scratch. Said the manager from a power plant in Shandong, "There was basically no special difficulty for us to apply for ISO 14001 certification, because we already had a relatively regular environmental management system before the application." However, areas of operation at a low level of environmental performance posed more difficulty than those already performing at a high level. "It is very difficult to substantially improve performance of some procedures, which started to operate from a very low level of environmental standards," he said. He also cited the continual improvement requirement as a particular challenge (Blue Sky Power Company environmental manager 2004).

An interesting concern cited by one environmental manager interviewed is rapid corporate growth and management change. The manager's Suzhou company grew from 1,800 employees to 4,000 during the year it obtained ISO 14001 certification. She said, "The only difficulty we faced is the relatively high frequency of change in company organizational structure, which created challenges to update all related documents in a timely way and maintain the EMS," (West Bridge Copy Machine Company (pseudonym) environmental manager 2004). Given the rapid pace of economic change in many parts of China, this may be an important issue for many enterprises.

On the positive side, two interviewees cited the creation of new income streams as EMS outcomes. The Shandong power plant environmental manager described coal ash recycling as increasing persistently through EMS implementation. It is sold for use in construction projects and earns the power company tens of millions of RMB in income annually (Blue Sky Power Company environmental manager 2004). Likewise, a Guangdong beer manufacturer earned 250,000 RMB in recycled glass and other waste in 2001 [Seafarer Beer Company (pseudonym) environmental manager 2002].

The case study reports we analyzed did not report any specific difficulties or disadvantages associated with certification. However, previous research on ISO 14001 implementation in China identifies some general issues that have been of concern to Chinese enterprises considering certification. For example, the problematic state of existing environmental regulation and enforcement in China has caused some firms to worry that pursuing compliance will get them involved in compliance matters that were not previously an issue (Morrison et al. 2000, p. 15). Additionally, firms are worried about the potential for conflicts of interests that arise when certifying organizations have the responsibility of both training enterprise staff on EMS management and ISO 14001 registration. Observers have also noted that there are difficulties in implementing the standard in older, state-owned factories and in small township and village enterprises, due to limited financial resources (Lam 1998).

5. China's ISO 14001 Adoption in Context

In the previous section of this contribution, we focused our discussion on understanding ISO 14001 certification at the enterprise level. Equally salient is the larger framework, or institutional setting, within which all aspects of enterprise environmental management— from the mandatory to the voluntary, from air emissions to wastewater discharges—plays out. One particularly important aspect influencing both the quantity and quality of certifications is that the various parties involved (e.g., transnational corporations, state government, and local government) hold differing perspectives on the purposes, costs, and benefits of certification. Another set of influential institutional factors involves how enterprises respond to mandatory environmental requirements.

National versus. Local Perspectives in ISO 14001

As described in Section 3, environmental regulators in Beijing have been very interested in the potential EMS holds as a sustainable development tool for China. The full set of "win-win" aspects of EMS implementation are well understood by actors at this level: enterprises can reduce costs and gain access to markets while lessening their environmental impacts. As such, ISO 14001 has been incorporated into a number of national policies and initiatives, which have shown some success at promoting ISO 14001 adoption.

However, it seems likely that, as with formal environmental policy, ISO 14001 is transformed in the devolution from Beijing to local authorities. Workshops on ISO 14001 for small and midsized enterprises (SMEs) have recently attracted minimal interest, suggesting that the benefits of an EMS may not be well understood by SMEs, or that they may not compare favorably to the other investments enterprises may be considering. The lack of interest in EMS among SMEs bodes poorly for ISO 14001 as a national policy tool, since SMEs account for 99% of all companies in China and 60% of the national economic output (China People's Daily 2002). The government also estimates that these enterprises make up more than 50% of its total industrial pollution load (CCICED 2003).

Likewise, the paucity of quantitative information about cost savings from EMS implementation begs many questions about the incentives we have identified for EMS adoption. Cost savings from EMS implementation usually result from reduced resource use, linking them more directly to environmental improvement than are trade access and corporate reputation, the other main financial incentives for EMS adoption. The latter benefits stem primarily from the ISO 14001 certificate, not directly from specific environmental improvements. If companies are not clearly quantifying and communicating their EMS-related cost savings, is decreasing resource use of less interest to enterprises than the benefits that come only from certification? If so we must question the likelihood of ISO 14001 to promote long-term environmental improvement for China.

Formal Rules, Informal Rules, and Incentives

On paper, the existing regulatory framework for managing industrial pollution in China is comprehensive, based in large part on the media-specific, command and control regulations developed by industrialized countries in the late 1960s and 1970s. In some cases, industrial standards for certain pollutants are even more stringent in China than in the U.S. (e.g., pharmaceutical wastewater COD discharges). Using water as an example, China has specific regulations on variables such as the amount of wastewater an enterprise is allowed to discharge in a given time period and allowable concentrations of key pollutants (e.g., COD). Enterprises that exceed these limits are required to pay fines or "fees" to the local environmental protection bureau. China also has a national water pollution law called the Three Synchonizations (S*an Tongshi*) that requires the design, construction, and operation of an enterprise's wastewater management infrastructure be conducted in tandem with its manufacturing facilities.

In practice, scholars studying the ground implementation of environmental policy in China have found compliance with these regulations to be a mixed bag of a success that Elizabeth Economy (2004) recently characterized as a "patchwork of effectiveness in implementation that depends upon the varying capacities, priorities, and incentives of local leaders." For example, Economy reports that only six of China's 27 largest cities supply drinking water that meets state standards (Wu et al. 1999), and that this problem is intensified by increasing amounts of industrial wastewater being generated by industrial and municipal activities.

Getting a handle on waste issues has proved a challenge to regulators. While compliance with state standards at select, targeted large enterprises is generally very good (Economy 2004, p. 70), very little is known about the enormous amount of waste being generated by the numerous, but for the most part, unregulated township and village enterprises (TVEs). This type of enterprise is quickly emerging as a major source of pollution problems (Smil 1997; Warwick 2003), yet the current setting for environmental regulation in China, as it is in many countries, is geared towards the source control of a few large polluters.

Economy (2004) describes long-standing institutional failures in the implementation of china's environmental policy from a national perspective. Beijing regulators, using successful Chinese economic policy as a model, have devolved authority to local- and province-level leaders. In many cases, incentives to enforce environmental regulations are weak compared to incentives to support industrial growth, and officials choose jobs over environmental protection. The economy's historical perspective is supported by the work of other researchers, including Sinkule and Ortolano (1995), Ma and Ortolano (2000), Cushing (1998) and others.

Taken together, previous research suggests a three-part framework that helps us understand how mandatory environmental policy implementation in China might conform the adoption of ISO 14001. The first leg of the framework concerns formal environmental regulations, such as those that limit an enterprise's wastewater discharges. Economy (2004), Ma and Ortolano (2000), and Cushing (1998) found in their research that environmental regulations handed down from above can be vague and even contradictory with other regional, industry, or local regulations. Local regulators must and do use their own discretion in determining exactly how to enforce them. State regulations are often disconnected from on-the-ground incentives, priorities and capabilities at the local level, where implementation and enforcement take place. For example, in her research on pharmaceutical factories in Shanghai and Shenyang, Cushing found that there was confusion in both the enterprise management's and local regulators' interpretation of the applicable standard for chemical oxygen demand (COD), as illustrated by the following quote from an engineering manager at a Shanghai pharmaceutical manufacturing plant:

"Our COD average for this year [1997] was 280 [mg/l]. However, it [the effluent COD concentration average] still doesn't meet the local standard of 100 [mg/l]. Our goal is to stay around the national standard, which is 250 or so. In reality, it is quite confusing. You have the national, general, and industry standards," (Animal Farm engineering services manager 1997).

The second prong of the framework is the set of informal rules, or what North (1990, p. 40) calls "informal constraints," unwritten but widely understood conventions of behavior and relationships that are extensions, elaborations, interpretations of formal rules (e.g., environmental regulations) that follow socially sanctioned norms of behavior. They are culturally embedded and internally enforced.

Informal rules play an important, and sometimes determinative, role in how enterprises respond to environmental regulations. These responses often differ markedly from the results intended by the original rulemakers. In their comprehensive study of over 228 enterprises across a variety of industrial sectors ranging from food processing to chemical manufacturing, Ma and Ortolano (2000) identified several key informal rules that dominated how certain environmental regulations, such as the Discharge Permit System, were carried out in practice: status or rank, *guanxi* (relationships), and *mian* (face). Below, we provide a definition and example of administrative status sample.

Status refers to the importance that both regulators and regulated entities place on the administrative rank, the rule being that regulators need to be of higher administrative rank than their regulatees if their orders or requests are to be taken seriously. In their research locales of Aryan, Changzhou, and Shunde, Ma and Ortolano (2000) found that local environmental protection departments were housed in departments that often had lower administrative rank than the enterprises in their jurisdiction making it difficult, if not impossible, for them to strictly enforce environmental regulations. Economy (2004) cites a similar situation in Guangzhou.

The final leg of the framework is incentives. The underlying theme is that, with appropriate inducements, economic or otherwise, enterprises can be motivated to achieve environmental performance improvement goals, especially in conjunction with local environmental improvement initiatives or cleaner production projects. Here, the work of Warren (1996), Cushing et al. (1999), and Economy (2004) is particularly instructive. In the case of Cushing et al. and Warren, enterprises showed that via cleaner production, enterprises could readily and successfully respond to rational, often low-risk economic opportunities that also reduced pollution. However, Cushing and coworker's (1999) research also points out that enterprise ability and the desire to meet ongoing or continuous environmental requirements (such as properly maintaining wastewater treatment equipment) or to take greater economic risks is often low. In her case studies, enterprises responded much better when the goals were finite, time-limited, such as with Three Synchronizations compliance. Finally, a number of researchers have talked about the effectiveness of local environmental protection campaigns as a tool for improving enterprise performance. Given the status quo informal rule that non-immediately threatening violations of standards are often tolerated, the presence or absence of local campaigns is often the factor that distinguishes between good- and poor-performing factories. Nevertheless, Economy (2004, p. 121) points out that the effectiveness of such campaigns can be limited due to their highly political nature, their tendency not to consult with local officials and businesspeople, and their tendency to not employ optimal technologies or incentives.

Applying Rules and Incentives to ISO 14001

Information presented earlier in our discussion demonstrates that market-based incentives to obtain "official" third-party ISO 14001 certification are high for organizations with certain characteristics (e.g., JV, supplier for TNC, electronics or telecommunications sector, pilot project participant), but institutional (e.g., regulatory) incentives to make substantive improvements in actual environmental performance are not as strong. So, while the *formal* rule governing ISO adoption might include actions such as fully addressing all significant environmental aspects, creating and using procedures and systems for monitoring, and striving for continuous improvement, the actual requirement for attaining certification is procedural. Given all the previous research demonstrating the importance of informal rules in determining organizational response to mandatory environmental regulations and the intentional vagueness of the ISO 14001 standard, it is highly likely that actual implementation of the standard across enterprises will vary dramatically.

At one extreme this could mean, for example, that a company could have a very progressive environmental policy on paper, but not actually carry it out. As stated earlier, Chinese enterprises do respond to finite, time-limited environmental requirements. Putting together all the paperwork required for obtaining ISO 14001 certification is a perfect example of this kind of task. However, the improvement model of an EMS requires that the organization continually work to monitor and better its existing system—an action akin to operating and maintaining wastewater treatment systems. For this kind of activity, Cushing's (1998) previous work suggests that enterprise incentive and ability may be much lower. Thus, the informal rule for the case of ISO 14001 may simply be to get certified by an external auditor, but not to truly integrate environmental aspects into all of the relevant components of the business or to strive for continuous improvement. Because violations of existing environmental regulations are often tolerated, the incentive to meet, let alone exceed, standards would seem to be low.

On the other hand, the right combination of factors, such as the market, strong and supportive enterprise leadership, and progressive transnational corporate governance could also lead to an exceptionally beneficial certification experience. With economic and managerial directives at the corporate and enterprise level, enterprises would have clear-cut incentives for addressing the more substantive aspects of the standards—aspects that are not necessarily part of the third-party certification process. This could be the case for manufacturing subsidiaries of transnational corporations with a strong reputation for good environmental management, such as Hewlett-Packard or Royal Dutch Shell.

A final application of previous research findings to the case of ISO 14001 concerns aligning incentives at all levels of implementation, particularly with regards to economics. As is the case with mandatory environmental regulations, there are significant differences in the way that various players in the ISO world view certification—regulators see it as another vehicle for enterprises to meet environmental protection standards; enterprises see it as a way to gain more international market share. As mentioned earlier, in our investigation we did not come across any significant evidence indicating clear net cost benefits to ISO 14001 certification. Thus, the often touted "win-win" benefits of improving environmental performance and saving money appear to be going unrealized in the firms we looked at.

The challenge for environmental regulators and policymakers is to better understand local level and enterprise incentives and design programs aimed at taking advantage of them. For example, Cushing and coworker's (1999) research indicated that firms were reluctant to invest in medium- and high-cost cleaner production options despite relatively quick payback periods. Further investigation revealed that during the training process, factory personnel were given very little training on how to conduct a detailed economic analysis. The researchers concluded that one explanation for the firms' reluctance was that decision-makers lacked confidence in the economic analysis. This points to a real need for increased training and education in environmental economics within the enterprises, perhaps sponsored and facilitated by organizations like SEPA, local economic and planning commissions, and industrial ministries.

CONCLUSION

In this final section, we consider the growth of ISO 14001 certifications in China and their ability to address what Economy (2004:103) describes as "the central tension inherent in China's environmental protection efforts: how to balance the development of China's economy with environmental protection, while maintaining the government's belief that less central government involvement is better." By taking into consideration factors driving enterprise adoption, available data on implementation results, important formal and informal rules that explain how enterprises manage their environmental issues, the realistic potential of this voluntary standard can be better understood. Finally, we propose ways in which the standard's efficacy can be enhanced.

If the current trend in ISO 14001 adoption continues at its current pace, by the end of the decade, China may lead the world in the number of certified enterprises. While Japan is still leading the charge in Asia, certifications are rising in many established Asian economies (i.e., Taiwan, South Korea, and Singapore), and Thailand appears to be following in hot pursuit.

Because of China's increasing regional leadership role, both economically and politically, Thai enterprises may look to China's implementation experience as an example of what to emulate in developing their own interpretations of the ISO 14001 standard.

Our research indicates that the effectiveness of the standard from an environmental standpoint has been mixed, and that the full economic benefits of implementation seem to be going unrealized. It also suggests that, at present, the informal rule of achieving *prima facie* compliance with the standard may be stronger than the stated formal rule of adopting a full-fledged continuous improvement-based EMS for many enterprises.

These findings, combined with the existence of a relatively weak environmental regulatory infrastructure and a nascent environmental nongovernmental organization (NGO) community suggest that continued ISO 14001 adoption, on its own, will not contribute significantly to improving environmental performance in Chinese enterprises. However, at a minimum, the standard helps promote environmental awareness within Chinese industry, which is the starting point for any kind of environmental performance improvement. The standard forces the enterprise to at least start thinking about and documenting how the main operations of its business impact on the environment. Thus, from an informational standpoint, certified enterprises are one step ahead of noncertified firms in this regard.

So, given that the number of ISO 14001 certifications will increase worldwide and particularly in Asia, what steps can be taken to maximize its potential effectiveness? We recommend two categories of action: 1) promote and support ISO 14001 adoption in ways that complement existing environmental regulations, especially those that have a market- or incentive-based component to them; and 2) take advantage of technical and economic expertise, as well as financial support, that can be provided by the international community.

ISO 14001 as a Complement to Existing Regulations

After numerous successful pilot projects, in January 2003 the Chinese government recently passed what may be the world's first law that established cleaner production, or pollution prevention, as a national strategy for achieving sustainable development. As cited earlier in the discussion, cleaner production programs help enterprise staff identify potential projects that result in quantifiable benefits, including cost saving, energy conservation, and improved product quality. The process of conducting a formal cleaner production project is very similar to establishing an EMS.

China's new Cleaner Production Promotion Law requires that new construction or significant renovations conduct environmental impact assessments with a priority placed on identifying cleaner production opportunities, specifically encouraging the use of recycled, recyclable, and toxin-free raw materials. It also has sections that relate to regional development planning that explicitly incorporates cleaner production, enterprise monitoring of resource use, encouraging environmentally responsible government procurement, and reward programs for exemplary enterprises (National People's Congress 2002). Cleaner production programs can provide the substantive base for an enterprise's EMS procedural framework and they also tap into the enterprise managers' increasing sensitivity to cost savings.

Another type of incentive-based environmental policy new to China is cap and trade, a form of emissions trading employed in many developed countries. Under such a program, the relevant government authority sets a "cap" on the total amount of allowable emissions from participating enterprises. It then creates tradable emission permits (in a unit of one ton of SO_2 per time period) that the enterprises can buy and sell amongst themselves. According to economic theory, the use of this kind of policy will result in the lowest total abatement cost. It also offers enterprises significant flexibility in determining compliance strategies.

In 2003, officials in the city of Taiyuan promulgated the Administrative Regulation for SO_2 Emission Trading, which authorized a cap and trade program involving 23 major sources. Because the program is so recent, no results are yet available, but analysts are optimistic that this law will be effective based on its detailed implementation requirements and strong penalties (e.g., high fines) for noncompliance. They also hope that the success of Taiyuan will form the basis for a national cap and trade emissions policy (Wang et al. 2004, p.36).

As with the nation's Cleaner Production Promotion law discussed in the preceding chapter by Mol and Liu, emissions trading taps into the increasing priority Chinese firms are placing on the economic bottom line. They provide direction for an enterprise's EMS. Strong enforcement of policies like these in conjunction with ISO 14001 may help align presently disconnected perceptions regarding the purpose of certification among auditors, regulators, and enterprises.

Help from Abroad

The history of China's interest in and adoption of the ISO 14001 standard is peppered with involvement from the international community. These organizations include multilateral lending institutions (e.g., the World Bank and the Asian Development Bank), international NGOs, the environmental protection agencies of other countries (e.g., the U.S. Environmental Protection Agency and Environment Canada), and TNC top management and technical experts. Ho (2001) points out that Chinese environmental activists are also heavily dependent on international donors as a source of funding, organizational assistance, and collaboration on environmental projects. Chinese leaders in all sectors of government have heartily embraced technical and financial assistance for environmental projects, particularly in the form of technology transfer-based pilot projects, which demonstrates their receptiveness to this kind of involvement. The future challenge lies not in creating successful pilot projects, but in moving to the next stage of diffusion where enterprises, on their own, will want to adopt EMSs for their own merits outside of the pilot project spotlight.

Topically, the subjects of environmental economics and project assessment would be a logical place to focus attention. By helping the enterprise staff develop economic assessment skills focused specifically on identifying linkages between a company's environmental and manufacturing activities, top company management will be better able to fully identify and have confidence in opportunities that concurrently save them money and improve the environment. Another approach to improving the effectiveness of international aid efforts would be to design programs that provide more enterprises with training, rather than focusing all project resources on only a few companies.

If combined properly with other, more substantive environmental regulations and environmental improvement initiatives that utilize all available domestic and international resources, ISO 14001 holds substantial promise. ISO certification can serve to support compliance with existing environmental regulations, thus assisting China's beleaguered

EPBs. ISO-certified entities have also demonstrated substantial improvements in areas such as air quality, wastewater treatment, hazardous waste management, and land use. Increasing ISO 14001 adoption, driven by international trade and supported by transnational corporate policy, can augment and complement China's existing environmental regulatory enforcement infrastructure, causing enterprises to voluntarily undertake what they would not otherwise engage in under mandatory regulation. As China moves toward incentive- and market-based environmental policies, such as the Cleaner Production Promotion Law and emissions trading programs, the potential for such improvements with ISO 14001 is likely to grow.

REFERENCES

Animal Farm (pseudonym), Engineering Services Manager, (1997), *Interview by Katherine Cushing*, 3 June, Shanghai, China.

Blue Sky Power Company (pseudonym), Environmental Manager, (2004), Personal communication with Hongyan Lu, 27 July.

Chen, Yanping, SEPA official, (2002), China Accreditation Committee for EMS Certification Bodies Presentation on ISO 14001 Implementation, 23 April, 2002, Beijing, China.

China Canada Cooperation Project in Cleaner Production (CCCPCP), (2001), "Implementation of the ISO 14000 System in the New Suzhou Industrial Park." Available: http://www.chinacp.com/eng/cpcasestudies/suzhou14.html, (Accessed 14 January 2004).

China Council for International Cooperation on Environment and Development (CCICED), (2003), "Recommendations to the Government of China," Available: http://www.harbour.sfu.ca/dlam/recommendations/2003%20recommen.html, (Accessed 14 August 2004).

China National Accreditation Board (CNAB), (2004), "Chinese Enterprise Certification Statistics," Available: http://www.cnab.org/cn. (Accessed 2 August 2004).

China National Accreditation Board for Certifiers (CNAB), (2004), ISO 14001 Implementation Statistics, Available: http://www.cnab.org.cn/, (Accessed 27 July 2004).

China People's Daily (English version), (2002), 20 September, Available: http://www.english1.people.com/cn/200209/20/eng20020920_103537.shtml, (Accessed 14 August 2004).

China People's Daily, (2000), "China to Fully Carry Out ISO 14000 Environmental Management System." 16 June.

China Registration Committee for Environmental Auditors (CRCEA), (2001), ISO 14001 Implementation Statistics, Available: http://www.naceca.org/crcea.htm, (Accessed 27 July 2004).

Copyer Company, Ltd., (2002), "Copyer Shenzhen Acquires ISO 14001 Certification." Press release from company website, Available: http://www.copyer.co.jp/e_topic/etpc_iso.html (Accessed 14 January 2004).

Cushing, K. K. (1998), "Wastewater Treatment and Cleaner Production in the Chinese Pharmaceutical Industry, How Institutions, Incentives, and Capabilities Influence Organizational Behavior," Ph.D. dissertation, Stanford University.

Cushing, K., P. Wise, and J. Hawes-Davis, (1999), "Evaluating the Implementation of Cleaner Production Demonstration Projects," *Environmental Impact Assessment Review.* Vol. 19, no. 5/6.

Geng, Y. and R. Cote. (2003), "Environmental Management Systems at the Industrial Park Level in China," *Environmental Management*, vol. 31, no. 6, pp. 784-794.

Hewlett-Packard (HP), (2004), "Environmental Management System," Available: http://www.hp.com/hpinfo/globalcitizenship/environment/operations/envmanagement.html (Accessed 16 August 2004).

Ho, P. and E. Vermeer (2006) (eds), *China's Limits to Growth: Prospects for Greening State and Society*, Oxford: Blackwell Publishers.

Ho, Peter. "Greening without Conflict? Environmentalism, Green NGOs and Civil Society in China", *Development and Change*, Vol. 32, 2001, pp. 893-921.

International Organization for Standardization (ISO), (1996), ISO 14001: Environmental Management Systems—Specification with Guidance for Use, ISO, Geneva, Switzerland.

International Organization for Standardization (ISO), (2002), *The ISO Survey of ISO 9000 and ISO 14000 Certificates, 12th ed.*, ISO, Geneva, Switzerland.

Kodak, (1999), Breakthrough Milestones in Chinese Imaging Industry, Press release from company website. Available: http://www.kodak.com.cn/CN/en/corp/pressCenter/pr19991111.shtml, (Accessed 14 January 2004).

Lam, Ying, (1998), "ISO 14000 Interest Gains Momentum in China," *IESU*, June.

Lamprecht, James L. (1997), *ISO 14000: Issues and Implementation Guidelines for Responsible Environmental Management*, American Management Association (AMACOM), New York.

Ma, X. and L. Ortolano, (2000), *Environmental Regulation in China, Institutions, Enforcement, and Compliance*, Rowman and Littlefield, Lanham, MD.

Matouk, M. (2000), "A Case-study of ISO 14001-based Environmental Management System Implementation in the People's Republic of China," *Local Environment*, vol. 5, no. 4, pp. 415-433. Nov.

Melnyk, S, R. Calantone, R. Hanfield, and R. Tummala. Center for Advanced Purchasing Studies (CAPS). (1999), *ISO 14001: Assessing Its Impact on Corporate Efficiency and Effectiveness*, CAPS, Tempe, AZ.

Morrison, J., K. Cushing, Z. Day, and J. Speir, (2000), *Managing a Better Environment: Opportunities and Obstacles for ISO 14001 in Public Policy and Commerce*, Pacific Institute, Oakland.

National Accreditation Center for Environmental Conformity Assessment (NACECA), 2001, "ISO 14001 Certifications by Region," Available: http://www.naceca.org, (Accessed 20 January 2004).

National People's Congress, Standing Committee, (2002), Cleaner Production Promotion Law. Available: http://www.chinacp.com/eng/cppolicystrategy/cp_law2002.html. (Accessed January 2004).

North, Douglass C. (1990), *Institutions, Institutional Change and Economic Performance*, Cambridge University Press, Cambridge.

Qian, Donglin and Li, Kaiming, South China Institute of Environmental Sciences, SEPA (2002). Interview with Heather McGray, 18 April, Guangzhou, China.

Ricoh Company, (2002), EMSs for Business Sites and Division, Information on company website. Available: http://www.ricoh.co.jp/ecology/e-/system/2.html. (Accessed 14 January 2004).

Seafarer Beer Corporation (pseudonym), former environmental manager, Interview with McGray, 20 April 20, Zhuhai, Guangdong.

Shenzhen Quality Certification Center EMS Auditor, (2002), Interview with Heather McGray, 20 April, Zhuhai, Guangdong.

Shenzhen Quality Certification Centre Corporate Brochure (English Version), (2002), March, Shenzhen Quality Certification Centre, Shenzhen.

Sinkule, B. and L. Ortolano, (1995), *Implementing Environmental Policy in China*, Praeger, Westport, CN.

Smil, Vaclav, (1997), "China Shoulders the Cost of Environmental Change," *Environment,* vol. 39, no. 9, July/Aug.

Spigelman, J. 2002, "China, rule of law or rule by law," *Quadrant*, vol. 46, no. 5, pp. 36-42.

Sustainable Business.com. (1999), "GM and Ford Require ISO 14001 from Suppliers," Available: http://www.sustainablebusiness.com/features/, 6 October, (Accessed 12 January 2004).

Teng, Jing, Director, Division of Industry Guidance and Accreditation, SEPA. (2003), Interview with Heather McGray, 16 January, Beijing, China.

United Nations Environment Programme-International Environmental Technology Center (UNEP-IETC). (2003), "Environmental Management Systems in China," UNEP, Osaka, Japan. Available: http://www.unep.or.jp/ietc/kms/data/910.pdf. (Accessed 13 February 2004).

Wang, Jinnan, Jintian Yang, Chazhong Ge, Dong Cao, and Jeremy Schreifels, (2004), "Controlling Sulfur Dioxide in China, Will Emissions Trading Work?" *Environment*, vol. 46, no. 5, June.

Warren, K. (1996). "Going Green in China, Pollution Prevention Practices in the Chinese Electroplating Industry," Ph.D. dissertation, Stanford University.

Warwick, Mara Kathryn. 2003. "Environmental information collection and enforcement at small-scale enterprises in Shanghai: the role of the bureaucracy, legislatures and citizens," Ph.D. dissertation, Stanford University.

West Bridge Copy Machine Company Environmental Manager, Suzhou, China. Personal communication with Hongyan Lu, 27 July 2004.

Wu, Changhua, C. Maurer, Yi Wang, Shouzheng Zue, and D. L. Davis, (1999), "Water Pollution and Human Health in China," *Sinosphere*, vol. 2, no. 3, p. 12.

Zeng, X.S. and H.C. Wang, 2002, "Determinants of Environmental Management Systems," Paper presented at the International Society for the Systems Science Conference on Systems Applications in Business and Industry, Shanghai, China, August 2-6, Available: http://www.systemicbusiness.org/digests/sabi2002/2002-206_Zeng_Wang.pdf, (Accessed 12 January 2004).

Zhang, Qichao, Vice Director, Shenzhen Environmental Management System Certification Center, (2002), Interview with Heather McGray, April 16, Shenzhen, China.

Zhao, Xiaodong, Chief of Administration and Public Relations, Center for Environmental Management and Certification Training, SEPA, (2002), Interview with Heather McGray, 22 April 22, Beijing, China.

In: Leapfrogging Development in Emerging Asia
Editor: Peter Ho, pp. 63-84
ISBN: 978-1-60456-410-5

Chapter 4

THE DEVELOPMENT OF ENVIRONMENTAL INDUSTRY IN CHINA: PITFALLS AND PROSPECTS

***Yi Liu*[1]*, Arthur P. J. Mol*[2] *and Jining Chen*[1]**
[1]Department of Environmental Science and Engineering, Tsinghua University, Beijing
[2]Department of Social Sciences, Wageningen University, Wageningen, The Netherlands.

1. INTRODUCTION

Profit-oriented industries and business activities are often argued to be at the source and origin of increasing resource depletion and environmental deterioration. In this respect, the *Environmental Industry (*abbr. EI[1]*)* can be seen as an exception. While it developed into a significant industrial sector in the 1990s, it emerged in the 1970s along with the increasing focus of industrialized societies on ecological and environmental issues. Instead of the resource consumption and environmental pollution of the traditional industrial sectors, the pollution control products and environmental services of environmental industries do not only bring economic benefits but also entail environmental gains, contributing to environmental compensation, rehabilitation, reuse and improvement. The EI is often seen as contributing to a path towards ecological modernization and greening (Mol and Sonnenfeld, 2000 and Ho and Vermeer, 2006) and sustainable development of human society on a macro-level and an efficient linkup between industrial productivity and environmental protection on a meso-level.

Unlike most conventional industries, EI firms fall under a wide range of conventional industrial classifications. This ambiguity makes it difficulty to identify and define whether

[1] "Environmental Industry" has been reviewed in different terminological ways worldwide. It is named as 'environmental protection industry' in China, as 'eco-business' in Japan, and as 'environmental industry' in US America. In other OECD countries EI is cited as 'environment industry' or 'eco-industry' frequently and seldomly as 'environmental goods and services industry.'

and to what extent a particular industry or firm belongs to EI or not. This is also valid for companies engaged in environmental consultation and services. That brings with it considerable difficulties in investigation, collecting industrial statistics and comparing EI development between different countries. Most countries adopt 'broad' definitions (such as the USA and Japan), which generally include a diversity of products and services related to environmental improvements during the overall lifespan of materials and products that could pollute the environment. The report 'Some Suggestions on the Development of the Environmental Protection Industry' issued in 1990 by the State Council indicated the first official recognition of EI as an individual industrial sector in Chinese economy. Originally, China adopted a 'narrow' definition, which took only into account products and services aiming at cleanup actions and remedial measures; thus a strong focus on 'end-of-pipe' technologies. Since 2000 it has been largely extended by different governmental departments, including the State Environmental Protection Administration (SEPA), the State Economic and Trade Commission (SETC, which was dismissed in 2003) and the National Development and Reform Commission (NDRC), toward an internationally accepted definition of EI. This shift, however, leads to data inconsistency and hampers the enabling of drawing comparative conclusions on the historical development trajectory of China's EI. More important, it seems unlikely that different governmental departments will reach a consensus of the definition of EI in the short term (see discussion in section 4).

After the Brundtland report (1987) and the United Nations Conference on Environment and Development (1992), EI grew rapidly and has been a large source of revenue in many developed countries. In the United States, EI is estimated to have had US$188.7 billion in sales in 1998, with 1,354,100 employees in 115,850 companies. This exceeds figures of several major manufacturing sectors, including chemicals, paper, and aerospace (Diener and Terkla, 2000). In the EU (EU15) EI has a turnover of approximately US$110 billion in 1998 and consists of 30,000 companies with 1,700,000 employees (ECOTEC, 1999).[2] Recently, the growth rate of EI in developed countries, in terms of sales, has slowed down somewhat.

China, as a developing country in transition from a centrally planned to a market-oriented economy, has been suffering severely from resource depletion and environmental deterioration. Especially in the 1990s, environmental problems in China have received increasing attention from governments and the public, both nationally and internationally. Some enlightened politicians, government officials and industrial managers in China have demonstrated their active concern for reducing pollution and encouraging cleaner production. Since the beginning of this century, an environmental campaign that aims at establishing a circular economy has been promoted by Chinese government. Furthermore, after 'environmental protection' was elevated to the status of 'National Fundamental Policy' in 1983 following the proposition of 'population control', 'resources saving' was newly proposed to be supplemented into the political regime at the fifth Plenary Session of the 16th Central Committee of the Chinese Communist Party (CCP) on 11 October, 2005, at Beijing. As a consequence, it can be reasonably expected that the huge domestic market for environmental products and services is growing. For instance, the annual construction cost of municipal water supply and wastewater treatment infrastructures was predicted to increase to

[2] Figures for 1999 are even higher: a turnover of 183 billion Euro and employment of 2 million full-time jobs (ECOTEC, 2002).

US$ 800-990 million in 2020 and gradually decrease to US$ 316-490 million until 2050[3]. An additional amount of US$ 6.0-6.2 billion and US$ 6.9-9.4 billion will be needed annually in terms of operational cost in 2020 and 2050, respectively (Chu, Chen and Zou, 2002).[4] In contrast to this 'huge' demand, and regardless of data inconsistency, however, China's EI has not been stimulated strongly by governmental regulation and public pressure, having only US$13.3 billion in sales in 2000. This represents less than 1% of China's gross domestic product (GDP) and only about 1% of the global sales volume (see Figure 1).[5] With a planned growth rate of about 15% each year within China's Tenth Five Years Development Plan ('*Shiwu Plan*'), and a prediction that gross production would achieve US$24.7 billion in sales in 2005 (SETC, 2001), China's EI could still be seen as relatively immature. The predicted 2005 production level would then consists of environmental equipment production (US$6.8 billion, 27.5%), comprehensive utilization (or reuse and recycling) of resources (US$11.7 billion, 47.5%), and environmental services (US$6.2 billion, 25.0%).

The purpose of this chapter is to analyze the strength and weaknesses in the further development of China's environmental industry towards maturation. Section 2 starts with a short review of China's EI, involving historical trends and the present status quo in comparison with global markets and domestic economic growth. Section 3 provides a discussion on some of the EI's developmental mechanisms, both from demand and from supply side perspectives, respectively. Because institutional constraints with respect to the administrative structure and environmental regulation form the key factors hampering EI's innovation in future China, it is given special attention in section 4. Conclusions with respect to the prospect of China's EI are drawn in the final section.

2. THE DEVELOPMENT OF CHINA'S ENVIRONMENTAL INDUSTRY

Although China's EI has made considerable progress in a few coastal cities in the last ten years (see section 3.3), in comparison with both EI's global development and the national GDP there is no evidence that indicates a significant overall innovation and development in this industrial sector. Figure 1 shows that although the absolute amount of firms and sales might look considerable, relatively it is far less than that of the United States or the EU (15) in 1998.[6] The exception here is the employment rates, which are of a comparable magnitude, showing especially the inefficiency of Chin's EI. The evolution of China's EI (illustrated in Figure 2) shows gradually increasing trends, indicated by the proportion of sales to GDP,

[3] An exchange rate of 1 US dollar for 8.1 RMB yuan is used in this chapter.

[4] Some other relevant data: 1) desulphurisation of 5% of the coal-burnt power plants in China could create a considerable market demand of RMB¥ 6 billion annually; 2) products for quieter and cleaner motor manufacturing could lead to more than RMB¥ 4 billion in sales each year; 3) while the construction of a domestic waste incinerator of 1,000 ton/day capacity costs about RMB¥ 0.6 billion, there are over 600 cities in China with a need but that have no waste incinerator. See details on the web site of http://sdep.cei.gov.cn/.

[5] In order to articulate a reasonable state-of-the-art of China's EI with respect to both historical trajectory and international comparison, the data derived from the China Environmental Statistical Report (various years) are used in sections 1, 2 and 3. Data inconsistency related to administrative institutions will be given special attention in section 4.

[6] EI definitions vary among these countries. Therefore, the comparative analysis should be interpreted as a qualitative illustration. Statistical data for industrial development derived from multiple sources show limited differentiation in magnitude (see Section 4). This fact justifies the conclusion on the status of China's EI, as drawn from the figure.

which increased from 0.22% in 1989 up to 0.87% in 2000. This is even more spectacular when taking into account the rapid economic growth over these years (average 9.4% annually).

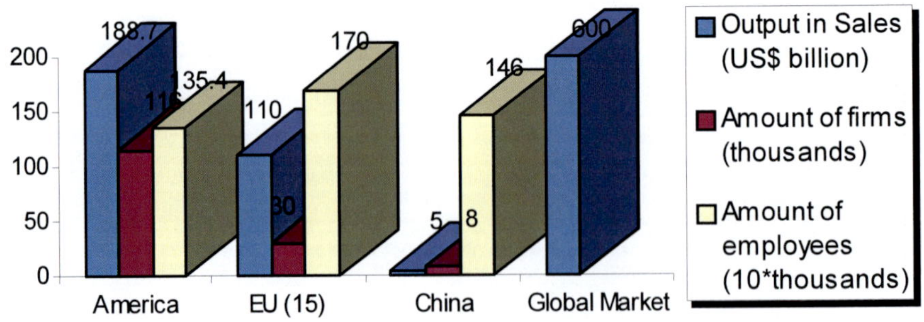

Source: Diener and Terkla, 2000; ECOTEC, 1999; the China Environment Statistical Report, 1998.

Figure 1. A comparison of EI's development among countries in 1998 and the global market in 2000.

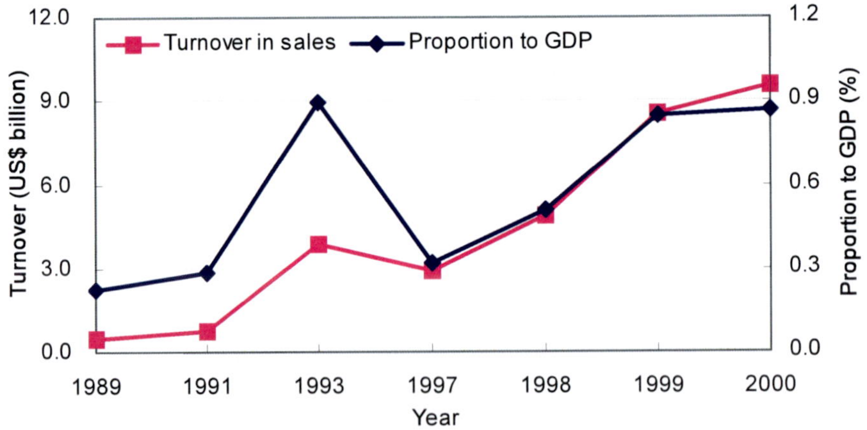

Source: The China Environment Statistical Report, the China Statistical Yearbook, 1991~2000.

Figure 2. EI's turnover in sales and proportion to GDP in relevant years.

This capital growth should not distract our attention from the relative underdeveloped status of China's EI. Several qualitative features indicate this underdevelopment. First, China's environmental industries are characterized by their small-scale and scattered distribution. The township and village enterprises (TVEs) account for a major share in this sector. Second, the quality of most environmental equipment, products and services is comparatively low. It is argued that the technological level of the China's EI is in principal comparable to the level of the western environmental industries during the 1960s and 1970s (Chen and Liu, 2001). Only a few companies can match current international standards. Repetitive investment in RandD for low-technology equipment, mass production of standard, low quality technologies, and even simple imitation, are prevalent among Chinese firms, particularly among TVEs. Third, incomplete and inadequate competition caused by an immature market mechanism and regional protectionism significantly undermines the innovative and RandD capability and capacity of Chinese EI, and results in a disadvantageous position in the global market. There is evidence that the national trade deficit of products related to environmental protection accounted for US$960 million in 2003 (Wang *et al.* 2004). Finally, most environmental companies and enterprises in China provide environmental equipment and products for end-of-pipe pollution control of industries, rather than focusing on ecological rehabilitation, environmental management and services of engineering projects, and information and consultation. To summarize, China's EI has stagnated at a preliminary stage in which its development is dominated by not very advanced 'end-of-pipe' technologies, strongly triggered by command-and-control regulation. Only marginal attention is paid to resource saving, energy efficiency, cleaner production, greening products and environmental services.

What is behind China's limited capability to develop EI further, both qualitatively and quantitatively? Broadly speaking four causes can be identified. First, the combination of weakened centralism and incomplete market liberalism, characteristic of the current transitional process in China towards a Socialist Market Economy,[7] brings with it compartmentalization of administration and regulation, market fragmentation, and regional protectionism (see section 4). Despite an increasing environmental policy capacity (see Figure 7) the implementation of environmental policies and standards falls short, leading to China's incapacity to stimulate market demands for EI (Fu, Chen and Zhang, 2004). Second, the current financial regime and practice of environmental protection in China is rather conventional, lacking sufficient and efficient fiscal and financial incentives. Governments have a central position in investment, construction, and operation of environmental infrastructures, but fail to commercialize this successfully (Chen et al., 2001; Fu, Chen and Chang, 2005). Third, China has poor capacity to commercialize scientific and technical RandD results at both national and industrial levels. The organizational structure of RandD in China fails to bring together research results of scientific and technological research institutions, environmental improvement demands of industrial firms or commercial companies, and state environmental regulation and policy. This significantly weakens the supply side: the effective and sufficient provision of advanced environmental equipment, products and services. China has not been very successful at applying domestic RandD results, nor at absorbing and assimilating imported environmental technologies (Alam, 2001). Fourth and finally, China misses a strong civil society articulating environmental interests, pressing the implementation of environmental policies and thus promoting market demand for the products of EI. The consultation and participation of environmental NGOs, communities and the public in environmental policy-making are still underdeveloped in China (Martinsons, 1997; Martens, 2006), while their freedom to widely circulate and publicly articulate environmental concerns is still restricted.

Besides ineffectiveness in dealing with environmental problems, the underdevelopment of environmental industry has also economic consequences. EI's underdevelopment results in large national pollution regulation and control costs and puts a heavy financial burden on

[7] 'Socialist Market Economy' is a specific concept that emerged in the 1980s in China and was formally defined in the early 1990s. That "a socialist economy can also be a market economy" is a historical and revolutionary breakthrough of traditional Marxist-based theory on socialist society. Learned from its own and other 'socialist societies' past experiences and lessons, China finally determined that 'establishing a socialist market economy' is the goal of economic reform. Consequently, market competition has been introduced into China's economy. This change has far-reaching impacts on Chinese society. Although it is still too early to grasp the final end-stage China is moving to, it is clear that China would have lost its 'global identity' if it had not taken the path to a market economy.

central and local governments. In addition, it is argued that the annual national wealth loss of China caused by environmental pollution and ecological deterioration accounts for 3.5~7.7 percent of GDP. Inefficient construction and poor management of urban infrastructure is responsible for 20~25 percent of this total loss (Chu and Chen, 2001).

It is apparent that EI's further development in China is strongly related to a successful transition, both in the economic and the political system. Under a market regime, it is possible to install a better incentive system to undertake technological and managerial environmental innovations, to stimulate adoption of advanced environmental technologies, and to promote the innovation of products and services. A more liberal political system will put stronger civil society pressure on reaching high levels of environmental quality and provide transparency on procedures, costs and information on the environment. Undoubtedly, EI's innovation and development in future China will be forced by four major mechanisms: 1) governmental rules, regulation and enforcement, with a focus on cleaner production and technologies; 2) continuous governmental investment and financing in public environmental infrastructure; 3) industrial polluters seeking for cost-reduction by lowering consumption of energy and materials and more (cost-)effective environmental technologies; 4) public environmental interest and demand for high living quality.

3. Demand and Supply for China's Environmental Industry

The innovation and development of environmental industries are determined by both the demand side and the supply side. Increased market demand lowers political and economic risks and market barriers for environmental industries, leading to large-scale environmental industries and contributing to national economic growth. An effective supply structure of environmental techniques, products and services may match the political and social requirement for environmental improvements.

3.1. Analyzing the Demand Side

Market demand for environmentally sound technologies, products and services is to be initialized and shaped by governmental promotion, industrial environmental performance requirements, and public pressure. The EI's market in China comprises: 1) governmental purchase programs or projects for ecological remediation and pollution control at the national and regional levels; 2) municipal infrastructure construction for public health and urban development; 3) industrial need for pollution abatement, cleaner production and resources or materials saving;[8] 4) household consumption focused at better physical health, comfort and environmental preferences; and 5) global market demand.

[8] Environmental (sound) technologies can be sharply split into control technologies and clean technologies. Normally control technologies do not require a significant redesign of the production process or products. The compatibility of clean technologies with existing systems of production is usually more limited. Thus, there is a relatively clear market demand for control technologies, while that for clean technologies is relatively vague and difficult to identify and estimate. See also Murphy and Gouldson (2000).

It is rather difficult to measure how large total market demand is with respect to these five clusters. Existing estimates mainly focus on the market for urban (water) infrastructure and industrial environmental improvement (such as those listed in section 1). These two clusters form also the major share of EI's market in China, due to the fact that urban infrastructure takes a priority within governmental programs, private environmental consumption has not been successfully motivated nationwide and international market demand is not very relevant for Chinese environmental industries.

Investments by the central and local governments, partly dependent on the national and local economic development, have played a significant role as driver and financial source for urban infrastructure construction and public sanitary improvement in China. As shown in Figure 3, notwithstanding the gradual increase of governmental environmental investment, the environmental investment per unit GDP has fluctuated and stagnated, at least until 1997. And even the highest level of 1.4 percent of GDP in 2004 is still lower than what can be witnessed in many industrialized countries.

Chu and Chen (2001) argue that, if the Chinese government maintains its existing financing pattern, the capital shortage for urban water treatment infrastructure construction would in ten years reach 30~60 percent of total needed capital, contributing to a considerable loss of national wealth. A healthy and sustainable development of China's EI depends on whether governments can successfully diversify the environmental capital sources in the next few years.

Environmental strategies and policies through governmental regulation and intervention (Gunningham and Sinclair, 1998) still construct the main drivers to improved industrial environmental performance. In this sense, how the government regulates polluting industries and to what extent it is successful has major consequences for EI's market demands on industrial point sources. Only in a few cases, however, endogenous environmental improvement arises in traditional industrial sectors in China (Vermeer, 1998; Ma and Ortolano, 2000; Mol, 2006; Liu 2005).

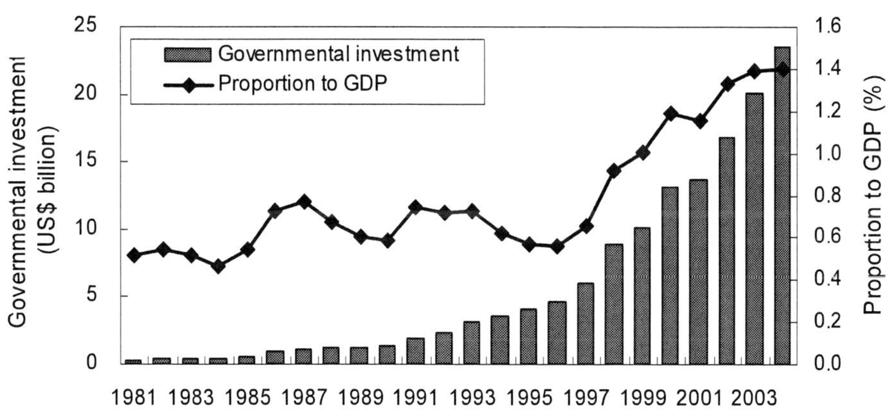

Source: Zhang, 1992; the China Statistical Yearbook 2004, the China Environmental Statistic Report 2000-2004.

Figure 3. Environmental investments by government in China, absolute (in million US$) and as proportion of GDP.

Regulation and legislation, but especially implementation and enforcement, are still too weak to foster EI's healthy growth in China (Lieberthal, 1995; Ma and Ortolano, 2000; Liu 2005). Too often the primary and only concern in China's political and administrative system is on economic growth (Jahiel, 1998). Local environmental protection bureaus, the main regulators of industrial pollution control and promoters of EI's growth, have to take economic growth concerns of local governments into account when regulating industries, since they are strongly dependent on local governments for budgetary funds and the allocation of administrative resources. At the same time, even under public pressure and governmental force, industries hesitate to improve their environmental performance, because environmental costs cannot be fully measured in advance, pollution charges are not fully returned to polluting industries that paid them, and environmental investments cannot bring immediate economic profits. Industries also fear that environmental measures will prove to be not as cost-effective as promised initially, as rules change regularly following policy fluctuations in a period of rapid transition. Most environmental officers in polluting industries focus on risk avoidance and security management, instead of creating environmental-economic opportunities.

Public consumption of environmentally-sound goods is hardly an effective factor in market demand for EI's innovation in contemporary China. While in the major urban centers green consumption is starting to become an issue among small groups with high cultural and financial capital (Martens, 2006), in the large majority of urban and rural areas and social strata limited environmental articulation in consumption does not trigger environmental industry development. According to an investigation in YunNan province in southwestern China, rural residents of five towns and seven villages surrounding the extensively eutrophicated Dianchi Lake were willing to pay an average of US$ 21 per household (RMB￥170; 1.7 percent of an annual total household income) for environmental improvement and ecological restoration of the lake.[9] While a few interviewees acknowledged a common duty among citizens to clean up the lake, 34.3 percent of the interviewees considered the governments responsible for Dianchi Lake cleanup and refused to pay anything for environmental restoration.

3.2. Supply Side Analysis

Chinese environmental industries supply different products and services. Chen and Liu (2001) have categorized these as follows: 1) environmentally sound technologies and related technological support; 2) environmental products and equipment including physical installations, debugging, operation and maintenance; 3) programs – consisting of proposals, planning, design, construction, operation and management – for ecological restoration and environmental improvement; 4) environmental information and consultation for industries,

[9] This investigation was conducted by experts and students of Tsinghua University (China) in 2001. In total 1024 households were interviewed to assess willingness to pay for environmental restoration. According to statistical analysis using contingent valuation method, estimated payment willingness averages US$ 32.6 (264.17 RMB Yuan) per household per year, with a sampling error of 3.5 percent within 95 percent confident interval. Note the difference between the average value of samples and the calculated value. For details see Wang and Zhang (2001).

public and governments; 5) environmental training, management and other services. It is not always easy to sharply distinguish these five categories in practice.

Research and development (RandD) capacity is crucial for the supply of environmental goods and services as it is one of the conditions for industrial technological capacity and market competency. But research and development capacity alone is not enough for an adequate supply structure. China's academic research on environmentally sound technologies can keep pace with that of developed countries in general, but product invention and design, product manufacturing, product quality, and systematic automatic control fall behind. Several causes are behind China's limited institutional infrastructure for environmental knowledge transfer and RandD application in both public and private sectors (Asuka-Zhang, 1999; Guerin, 2001). First, in China technological innovation and innovative research take place primarily in academic institutes and public sectors, but hardly in industries themselves. Second, academic RandD results do not match industrial demands and are often unsuitable for large-scale commercial application, leading to low levels of commercialization and utilization of environmental RandD results. Third, innovative capabilities and capacities are predominantly geared towards absorbing, adapting and duplicating technologies already developed in industrialized countries. Little progress on novel environmentally sound technologies has been made in China up to date. Fourth, this tendency is further strengthened by demand side factors: polluting industries prefer overseas environmental technologies, even in cases where domestic ones have lower prices, are easier to maintain and have comparable quality. Most of the polluting industries consider foreign products superior in quality and performance and find foreign suppliers more reliable and experienced.

And finally there are financial reasons behind the shortfall in developments at the supply side, to be broken down as follows (Preston, 1997):

- Academic environmental RandD lacks sufficient governmental subsidies;
- polluters are charged with very low fees, which undermines the incentives for industrial environmental innovation;[10]
- lack of funding weakens the environmental RandD undertaken in industries;
- relative high market risk for suppliers, also due to regulatory failure. Most 'technology-based regulation' focuses on 'choosing' existing technologies, rather than encouraging 'technology development';[11]
- little interest of private investors and capital suppliers in environmental business, because of large capital amount needed, long paybacks periods, high risks, and regulatory uncertainty. The gap between the government and market in the commercialization of environmental technology is too large to bridge. It commonly arises after the fact, that government officials conclude the project to be too commercial to fund and before investors are willing to make risky investments.

---

[10] This is also an important factor influencing industries' decision on environmental performance. See Ma and Ortolano (2000) for an empirical study and rich discussion on the gap that exists between the goals embodied in China's environmental laws and regulations and actual levels of environmental quality.

[11] There has been a remarkable shift in the technology polices of developed countries. Governments are increasingly seen as facilitators rather than providers of direct support to innovative activities. Consequently, while governments continue to provide fiscal support to R&D, the focus of policy has shifted to strengthening the interfirm and academic-industrial linkages. See details in Alam (2001).

Global suppliers' impact in China was already large in the late 1990s (see Table 1), but is further strengthened after China became a member of the World Trade Organization (WTO) in 2001.

Table 1. Market size and major supplier shares in 1999

Market	Market size (billion Euro)	Domestic share (%)	US share (%)	Japan share (%)	EU share (%)
North America	132	96	-	1	2
EU	104	97	2	1	-
Japan	62	97	2	-	1
SE Asia	9	48	20	15	15
China	4	55	15	15	12
South America	4	44	30	5	18
CEE	4	27	20	3	45
Australia/NZ	4	59	25	10	5

Source: ECOTEC (2002).

Although lower barriers are beneficial for China's EI in terms of foreign capital, advanced technologies and management experiences, the parallel severe international competition pushes national industries out of domestic markets. China's water treatment systems, for instance, have been plagued by debt and low environmental efficiency under state-owned enterprises. The French company Suez Lyonnaise Des Eaux, one of the world's biggest suppliers of water treatment technologies and services, is increasingly active in contemporary Chinese water treatment programs, setting up nine jointly owned water treatment plants in China recently.[13] These and other foreign investments bring modern technology and equipment to China, as well as efficient management that help to reduce costs and increase profits. If the immature domestic environmental industry cannot compete with the new entrants in the Chinese market, China might also lose relatively low cost domestic pollution control in the end.

3.3. Regional Differentiation

Although in general China's EI is still immature, environmental industries in a few economic regions have made considerable progress. In Shanghai city, the most energetic area in China, the environmental industrial system consists of 630 domestic and foreign enterprises and nearly 100 research institutions, with environmental investments of 2.8

[13] *China Daily*, March 29, 2001. In Chengdu, the capital of Southwest China's Sichuan province, a foreign-owned water project called Vivendi Waterworks will begin supplying tap water in March 2002. The France-based Vivendi Water, one of the world's top companies in this field, set up Vivendi Chengdu Waterworks Co Ltd for water supply (*China Daily*, August 10, 2000).

percent of total GDP in 1996-2000, which ranks 1.8 points higher than the national figure.[14] In Shenzhen in Southern China, EI's sales achieved 318.8 million dollars in 2000 (23 times higher than sales in 1997), while local GDP reached 20.6 billion dollars in the same year (Shenzhen EPB, 2001).

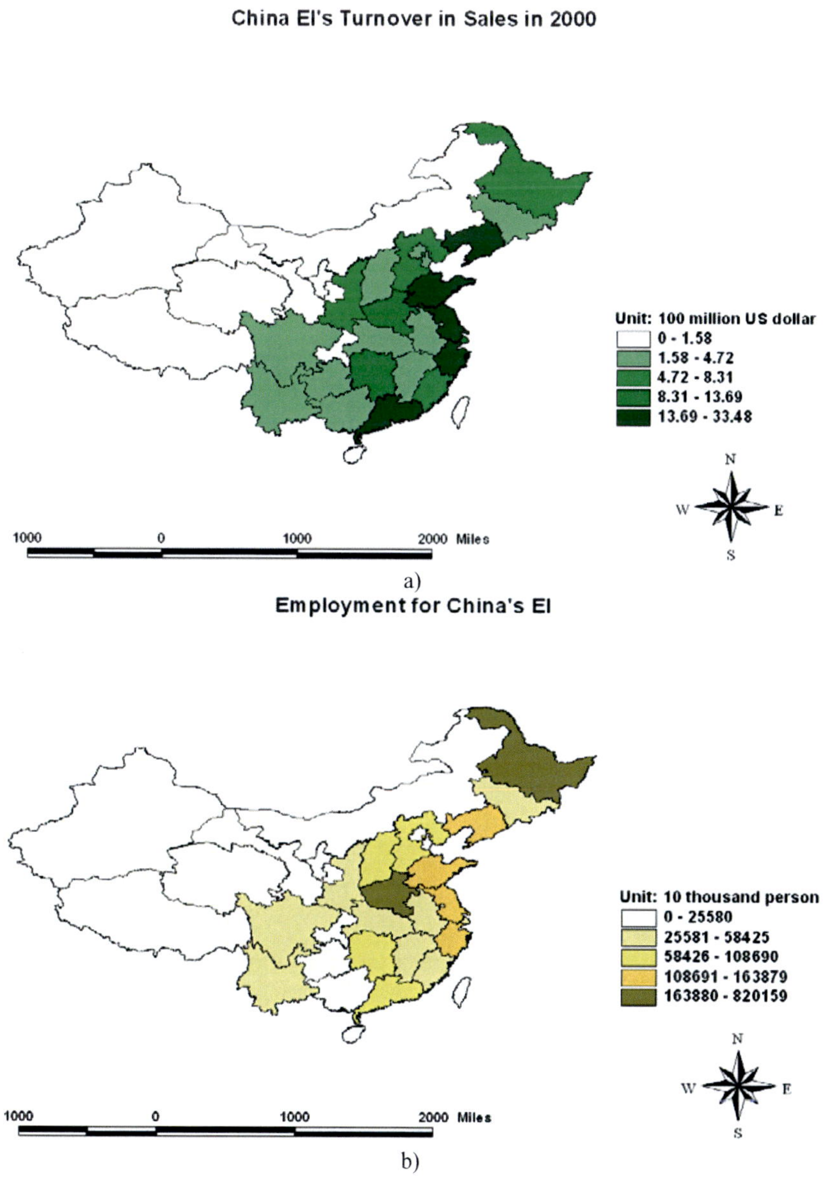

Source: SEPA, 2001.

Figure 4. Regional development of China's EI in 2000: a) turnover in sales; b) employment.

---

[14] Shanghai aims to evolve its environmental protection sector into a principal industrial sector by 2015 with a total output of US$12.3 billion. Sales in environmental protection facilities and equipment are expected to reach US$1.85 billion in 2000 and US$4.9 billion in 2005, with an annual growth rate of 21.6 percent. Output is expected to amount to US$1.1 billion in 2000 and US$2.5 billion in 2005, taking a 25 percent share of the domestic market. ('Profits make for a greener city,' *China Daily*, 2 Sept. 2000).

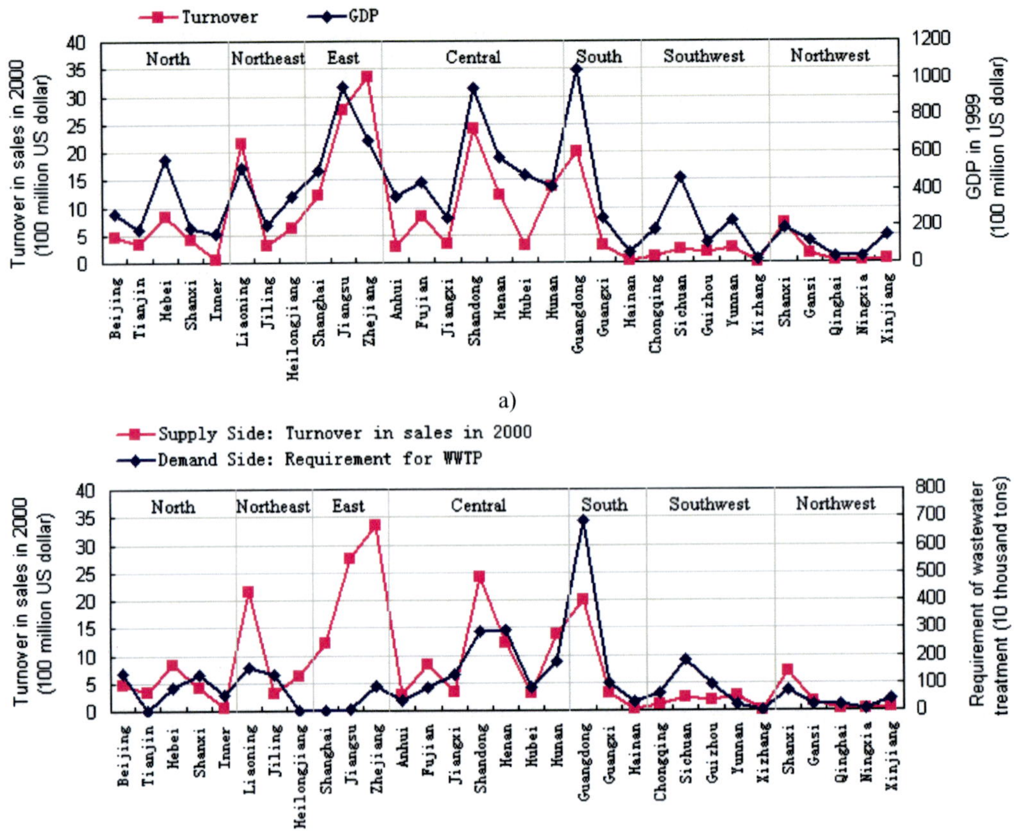

b)

Source: SEPA, 2001; the China Statistical Yearbook, 2000.

Figure 5. Regional development of China's EI: comparisons of market supply in terms of turnover in sales with: a) local GDP; b) market demand for wastewater treatment.

The geographically uneven development of China's EI is shown in Figure 4. EI's sales indicate an uneven supply of environmental technology and services, ranging from high amounts in the southeast to much lower ones in the northwest. Regionally uneven economic development parallels the unbalance of China's EI, which is further illustrated in Figure 5a. This is not surprising because local economic potential and capital availability, among others, stimulate the development of environmental businesses in local areas, as well as the application of their goods and services. While the overall regional distribution of employment in environmental industry is consistent with that of its sales, Figure 4b illustrates that in some inland areas, such as Heilongjiang and Henan provinces, market supplies go together with relatively high employment figures, while other provinces, such as Guizhou and Guangxi Zhuang, show low employment rates with relatively high sales.

There are also strong local differences in the 'gaps' between supply and demand. Using the shortage of urban wastewater treatment to represent market demand,[15] market demand and supply are compared at local level (as illustrated in Figure 5b). In some regions, such as Shanghai, Jiangsu, Zhejiang and Liaoning provinces, EI's supply seems to be sufficient to

[15] Calculated by the difference between the volumes of urban wastewater discharge and the treatment capacity in 1999.

meet the demand for urban wastewater treatment. But in some other provinces, involving Guangdong, Shangdong and Henan, notwithstanding the relative high market supply of local environmental industries, there exist still large gaps between present treatment capacity and the volumes of urban wastewater discharged. Note that overall the limited supply of environmental industries in the northwest region is sufficient to meet wastewater treatment demand. It goes without saying that in each of these regions and provinces the averages hide local situations with major shortcomings in treatment capacity.

4. Regulating Environmental Industry in China

Since environmental regulation is fundamentally related to both demand for and supply of EI's products and services, it is interesting to analyze the environmental state's institutions for environmental improvement and pollution control in relation to the development of EI.

4.1. The Environmental Matrix Problem

The environmental regulatory administrative system is organized as a hierarchical system: from the national level, via the provincial, the municipal, and the county level, to the township level.[16] The administrative (or bureaucratic) rank is an important reflection of power and status. The highest-ranking units under the State Council are the comprehensive Commissions, which rank higher than ministries. At the township level and above, governmental organs are connected in two distinct ways: by function and by geographical area. The Chinese use vivid terminology to describe these criss-crossing jurisdictions: the vertical bureaucracies are called lines (*tiao-tiao*) and the horizontal coordination at various levels is called pieces (*kuai-kuai*). Western scholars of organizational dynamics label this dualism of horizontal and vertical authority a 'matrix' problem (Lieberthal, 1995: 169). In the complex *Tiao-Kuai* (matrix) relationships each unit reports to both an upper-level department in the same functional area and the government of the geographical area in which it is located.

As the major advocator and promoter for environmental interests and EI's innovation, SEPA plays a key role in designing pollution control policies and related programs. SEPA was elevated in the fourth administrative reforms in March 1998,[17] from vice-ministerial to ministerial ranking, and the amount of employees in the SEPA line (*tiao*) nationwide has almost doubled in a decade: from some seventy thousand in 1991 to over one hundred and

[16] Governments at the township level and above are 'official' in the following sense: they receive budgetary support from the Centre and their employees are regarded as formal governmental staff. In contrast, the governing body in a village is called 'village committee' which consists of a leader, a deputy leader, an accountant, and so forth. See details in Ma and Ortolano (2000: 34). In fact, two bureaucratic hierarchical systems exist: the territorial one (center-province-city-county-township) and the administrative one (state-ministry-bureau-department-division). The relations between these two hierarchical systems are in China referred to as: *shengji-buji*, *tingji-juji*, *diji-chuji*, and *xianji-keji* (see Figure 6). To complicate matters further, there is a complex hierarchical relation between the state government and the communist party at each of these levels (e.g., Lieberthal, 1995).

[17] The Ninth National People's Congress (NPC) turned out to provoke a radical reform of government administration. When the dust had settled, the number of ministry-level bodies had been reduced from 40 to 29, and 50 percent of the governmental position had been listed for removal from the governmental payroll within three years.

sixty thousand in 2004 (see Figure 6). But SEPA's role in day-to-day implementation of environmental regulations is still limited. SEPA implements rules only for projects undertaken by national-level agencies, or activities that are otherwise of national significance, such as the famous 'three rivers and three lakes' project. In all other cases, Environmental Protection Bureaus (EPBs) or Environmental Protection Offices (EPOs), replacing SEPA through successively lower levels of the administrative hierarchy, enforce environmental laws, implement policies designed by SEPA and assist in drafting local regulations to supplement central ones (Ma and Ortolano, 2000: 9; Vermeer, 1998). Higher-ranking organs provide lower ranking ones with policy directives and guidance for implementation, but normally it is the local government that provides corresponding affiliated organs with annual budgets and funds and approves institutional advancements. Therefore, the local government is usually more powerful in directing EPBs and EPOs than SEPA: the '*tiao*' serves the '*kuai.*' Local government budgetary control of regulatory agencies suggests that local governments may have a larger influence on how policies are implemented than central-level regulatory agencies responsible for formulating policy (Jahiel, 1998; Liu, 2005). To summarize, the central government's ability to coordinate policy across regions may be seriously limited, which in turn significantly hampers the demand side of EI in China.

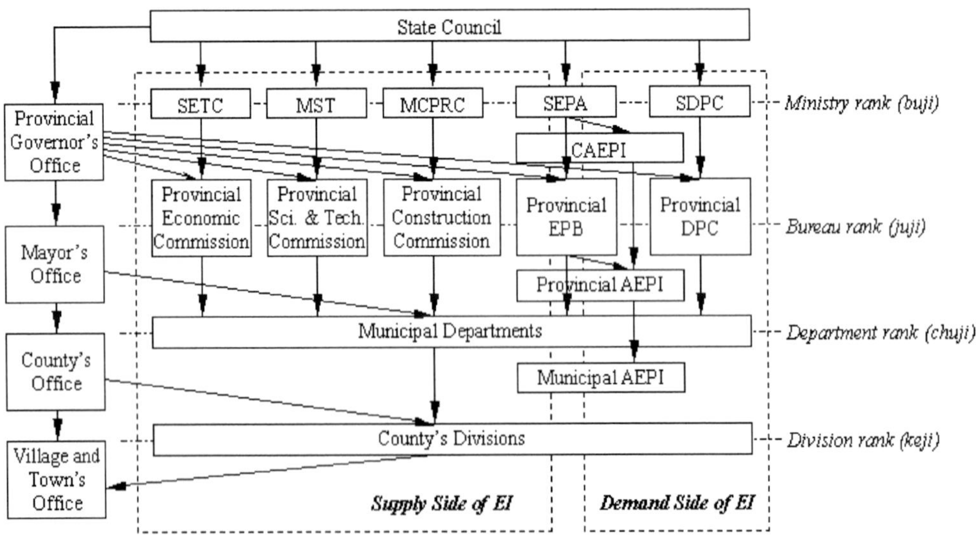

Source: The China Environment Statistical Report, 1991~2004.

Figure 6. Amount of staff employed for environmental protection in years-end in China.

In 1993 SEPA set up the China Association of Environmental Protection Industry (CAEPI), apart from the bureaucratic hierarchy but with local branches, in order to accelerate the growth of China's EI. In almost ten years CAEPI has made various contributions to the development and innovation of environmental industry, among which three surveys on EI's development in 1993, 1997 and 2000 respectively.[18] Although it has a semiofficial position, CAEPI is excluded from the policy-making bureaucracy.

[18] The major goal of CAEPI is to actively participate in governmental regulation for EI's innovation and development. In practice, this includes among others: 1) taking part in environmental industrial survey, strategy planning, and developing policies; 2) studying and reporting on EI's strategy and policy; 3) providing

Without administrative ranking CAEPI has no administrative privilege and governmental policy instruments to press and promote EI's innovation at the national or local level. Although it has received some 'administrative' regulation power and budget from SEPA, it can only wander in the complex and confused matrix to try to set favorable conditions for the development of environmental industries.

Concerning EI's innovation and development there are still several other agencies that have jurisdiction over various aspects of environmental protection besides SEPA. These involve the State Development and Planning Commission (SDPC, until 2003) and the State Economic and Trade Commission (SETC, until 2003), which have responsibilities that cut across sectors and geographic regions and contribute to the supply side and demand side, respectively. The other relevant units at the national level are the Ministry of Science and Technology (MST; responsible for the innovation and supply of environmental sound technology), and the Ministry of Construction People's Republic of China (MCPRC; responsible for urban infrastructure and environmental facilities). While the 1998 administrative reforms reduced the number of agencies involved in environmental protection, it did not completely eliminate the authoritative fragmentation. All these governmental agencies comprise a complicated administrative hierarchy related to regulating EI from the Centre to local levels, as illustrated in Figure 7. Some improvement was made following the fifth administrative reform in 2003, where SETC and SDPC merged into the powerful National Development and Reform Commission (NDRC), including the administrative functions on EI. But the matrix still keeps muddled. This complex arrangement of diffuse organizational tasks and responsibilities for environment and EI has several notable implications. First, not all responsible organizations are strongly inclined towards environmental interests as a prime motive. Second, this fragmentation creates competition for scarce environmental funds and for authority (and thus power). Finally, fragmented, and in some cases overlapping, regulatory authority hinders the coordinated management of environmental issues as well as EI's innovation.

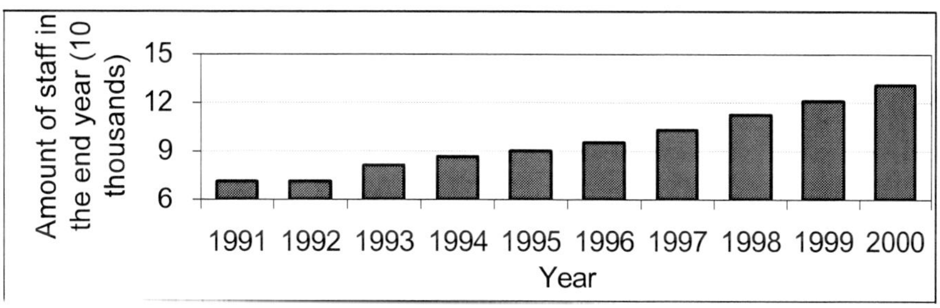

Figure 7. Regulating EI in China: the framework of administrative hierarchy (before 2003).

necessary information and consultation on environmentally sound technologies, equipments and services for polluters and suppliers; 4) assisting environmental industries with respect to technological, financial, and facility aspects. See http://www.beijingwindow.com/jj/hbcy/cyxh/.

[20] The details were promulgated in *A Complete Volume of Laws and Regulations of China's Environmental Protection: 1997~1999*, edit by Department of Policy, Laws and Regulations, SEPA, Xueyuan Publisher, Beijing, 1999.

4.2. Regulating China's EI: Hierarchical Ranking Versus 'Orbits'

A remarkable institutional transition is the fact that SEPA's jurisdiction on the development of EI was passed to SETC in the 1998 administrative reforms.[20] This regulatory restructuring implies the recognition of the significance of EI for the national economy.[21] SETC and the comparable organizations responsible for economic development and trade at lower levels have a higher administrative ranking than SEPA and EPBs. However, it seems that SEPA does not completely comply with this institutional adjustment. Two pieces of evidence are presented to understand the institutional overlap and conflict caused by the restructuring: 1) after the reforms an important inner department, called the Department of Technological Policy and Industry Direction within the Technology and Standard Bureau in SEPA, remains responsible for EI's development by taking a serious account of technological innovation of EI suppliers rather than only the demand side; 2) SEPA initiated a comprehensive investigation on EI's development at the national level in February 2001 (carried out by CEAPI) and came to conclusions (in the form of an internal report) that are considerably different from the ones in a report SETC (2001) launched in the same year.[22]

Three differences can be noted between the CAEPI/SEPA report and the SETC study: the differences in the label for EI, discriminating definitions of China's EI and (consequently) deviations in conclusions. Even when taking into account the diverse terminology used to present EI worldwide (see Footnote 1), differences in labels cannot be ignored in China because they are grounded in the conflict and contest between SETC and SEPA for regulatory domination. Following definitions of EI that are claimed to be more consistent and comparable with those used in Western countries, SEPA applies a broader definition than SETC, extending EI's category into five aspects: environmental protection products, clean products, environmental protection services, waste treatment and disposal, and nature and ecology protection. SETC still adopts the prevailing narrow definition, including three categories: environmental equipment, environmental services and comprehensive resource utilization (i.e. waste re-mining and recycling). No coordination or attempts at harmonization takes place between these two organizations. In addition, SETC statistical data rely on different sampling techniques and survey methods. Even when we exclude the two supplemented categories of SEPA's definition (clean products and nature and ecology protection), the statistical data of EI's sales in 2000 differ considerably between the SEPA and SETC reports, especially regarding environmental services and comprehensive resource utilization (see Figure 8).

[21] But there might be more at stake. SETC, and not SEPA, also managed to become the first responsible agency for, among others, the Cleaner Production Promotion Law. See Mol and Liu (2005).

[22] The internal report, named the State of the Environmental Protection Relevant Industry Report in China, has an appendix in which the deviation in acknowledgements of EI's status are emphasized and interpreted. Apparently, SEPA tries to highlight the significance of the report and advocate the 'novel' notions by comparing its study with SETC's 'out-of-date' one. It implies that SEPA is actively striving for the regulatory domination that was weakened in the administrative reforms. However, this last section of the internal report was excluded when it was circulated to the wider public, under the same title but with a different content, in December 2001.

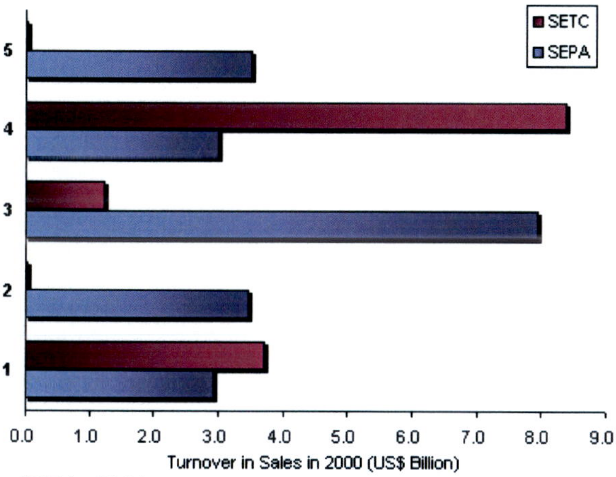

	SETC	SEPA
1	Environmental equipments	Envionmental protection products
2	/	Clean products
3	Environmental service	Environmental service
4	Resources comprehensive utilization	Waste treatment and re-use
5	/	Nature and ecology protection

Source: SEPA, 2001.

Figure 8. Statistical inconsistency between SETC and SEPA in terms of EI's turnover in sales in 2000.

Apart from SEPA and SETC, there is a third data source related to the development of EI: the China Environment Statistical Report (CESR). The CESR is promulgated by SEPA annually to proclaim fundamental statistical information on China's environment, normally involving pollutants loading and control, natural ecology protection, environment regulation and management, capacity building, and so forth.[23] A comparison of the three sources makes clear that EI's total turnover, amount of units and employees derived from the SEPA investigation are much higher than those derived from the other two investigations, as presented in Figure 9. If the broader SEPA definition is appropriate, it suggests that the development of China's EI has been underestimated over a long period.

The implications in terms of administrative conflicts involving overlap and fragmentation, which go behind these data sets, are more interesting than a quantitative analysis of data uncertainty. Depending on CESR and its subordinate affiliations, SEPA can be considered to play a dominant role in the EI regime development. With a weakened regulatory function after 1998, SEPA still conducts EI's investigations, promulgates the development EI to other governmental agencies and the public and tries to exclude any participation of SETC in this. The fact that a higher-ranking organ, SETC, cannot always intervene into an assignment organized by a lower-ranking one, SEPA, or rearrange and incorporate a program with it, strongly suggests that – besides ranking – another critical force defines administrative jurisdiction in the huddled matrix. There exist 'legitimate' privileges in the conventional '*orbit*' or '*perisphere*' occupied by each governmental organ, which are strongly institutionalized in the customs of Chinese policy culture and historically developed bureaucracy. In this case of distinct representations and positions on China's EI, SETC barely manages to intervene into SEPA's conventional and institutionalized environmental routines and domains, even though the former has a higher rank designated by the Center.

[23] While SEPA claims their industrial investigation reflects the actual status of China's EI, with an elaborate explanation of the extended categories in its report, the data series abstracted from CESR are adopted consistently in this contribution, since that seems to be in a better order and more appropriate for temporal dynamic analysis (see Figures 1, 2 and 6). On the other hand, whatever data series is used for comparative analysis as discussed in sections 1 and 2, the conclusion will most likely be the same: China's EI is relatively immature.

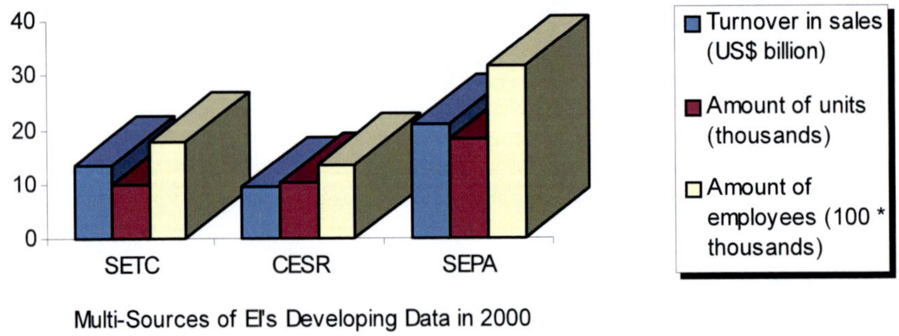

Source: SEPA, 2001; the China Environment Statistical Report, 2000.

Figure 9. Comparison of data derived from multiple sources on EI development in China in 2000.

On the other hand, however, SEPA's 'conventional orbit' cannot fully prevent SETC from taking forward steps on regulating EI (and other environmental policy domains). SETC has its own data sources and relies on its own definition in assessing EI's development. For SETC it is important that it can also promulgate the status of developing EI in China in the form of its own documents, despite data gaps. After the 'Encouraged Catalogue of Industries, Products and Technologies' issued by SDPC in 1997, in which the environmental subcategory was included, SETC documented the 'Encouraged Catalogue of Equipment (Products) of Environmental Protection Industry by the Government (the First Batch)' in 2000 (*Guojingmaoziyuan* [2000]-159) and supplemented the catalogue in the form of the second batch in 2002.[24] This can be viewed as an important step in the direction of effective policy-making on environmental (technological) industry. In 2001, a formal document, named 'Suggestion on Acceleration of the Development of the Environmental Protection Industry in China' (*Guojingmoziyuan* [2001]-517), was written by SETC and jointly issued by eight ministries and commissions. [25] The document claimed that – by its own words: different from SEPA – EI is a technological guarantee and physical base for environmental protection undertakings and an increasingly significant contributor to national economic development, and promotion of EI is a crucial measure towards sustainable development.[26]

[24] The classification of these two catalogues adopts the same taxonomy. It includes eight sections: 1) air pollution control; 2) water pollution control; 3) waste disposal and treatment; 4) noise control; 5) environmental monitoring; 6) energy saving and renewable energy utilization; 7) comprehensive utilization of resources and cleaner production; and 8) environmentally friendly materials and medicament.

[25] The eight units include SETC, SDPC, MST, the Ministry of Finance People's Republic of China (MFPRC), MCPRC, The People's Bank of China (PBC), the State Administration of Taxation of China (SATC), and the State General Administration of the People's Republic of China for Quality Supervision and Inspection and Quarantine (SGAQSIQ). Note that the last unit was newly established on April 16, 2001, which is actually a merger of two former government departments, namely, the State Administration of Quality Supervision of the People's Republic of China, and the State Administration for Entry-Exit Inspection and Quarantine of the People's Republic of China. SEPA was not among the eight.

[26] The document mentioned three key domains supported by central government: 1) environmental technology, equipments, environmental materials and medicament; 2) comprehensive utilization of resources; and 3) environmental consultation and information, technological services, and operational service of environmental projects and infrastructure. These three categories are consistent with those of the definition advocated by SETC. On the other hand, it is clear that SETC is striving for upgrading EI's position in the whole national economy. For example, the vice-director of SETC said that the National Planning of Ecological Construction and Environmental Protection in China's the Tenth Five Years Development Planning ('*Shiwu Planning*') treats EI as a novel and raising industry and it will certainly and significantly contribute to the national

SETC strongly recommended in the document that EI should be treated as a dominant sector, promoted by governments in the local economy as well as in social development planning and future goals.

According to the fifth administrative reform in 2003, SETC was dismissed and its function on EI was transferred to a newly (re-)organized agency, i.e., the National Development and Reform Commission (NDRC). As the highest ranking commission in administrative hierarchy, NDRC was assigned the responsibility, among others, for relevant laws, policies, plans and standards aiming to promote EI's development, cleaner production, and circular economy. This strongly suggests that the administration upon EI is further updated at the central level in parallel with the organizational rearrangement. It also can be expected that EI will play a more active role in the national environmental campaign of circular economy.

The latest administrative reform, however, did not touch the problem of institutional conflict in relation to regulating EI between SEPA and NDRC. SEPA and its affiliated CEAPI have been engaged in investigating and releasing information on EI. The data inconsistency still remains. For instance, SEPA presented that the turnover in sale of EI reached US$ 28.4 billion in 2003 (Wang *et al.* 2004), while the NDRC estimated it will be US$ 24.7 billion in 2005.[27] It is important to keep in mind that behind the data gap, the conflict, involving fragmentation and overlapping, remains inherent in the muddled matrix. This continues to influence both demand and supply sides for EI's innovation in China.

In general, China has not built up an integrated and highly efficient regulatory and promotional system for EI's innovation and development, nor an effective policy-making and implementation on both the demand and supply sides. The distinction between the administrative hierarchy and ranking (SETC/NDRC) on the one hand and the conventional 'orbit' (SEPA) on the other constitutes a major barrier in coherent and consistent regulatory jurisdiction, environmental reports, data collection and promotional policy.

Conclusion

The market and regulatory potentials for promoting EI's innovation in relation to both demand and supply have taken a central place in this contribution. With respect to the demand side, the emphasis should be on the possibility for ecological-economic 'win-win' solutions, by changing technical progress into a more ecological-efficient direction that also contributes to the overall competitiveness of Chinese industry. Although regulation can be used to force the process of industrial eco-innovation, regulatory intervention cannot completely overcome the institutional and structural constraints environmental industries are facing in their development. Even if environmental regulation can trigger more than incidentally incremental or radical innovation in polluting industries, there may in the end be institutional, political and economic constraints, particularly in transitional China, which make continuing

economy by the central government's stimulation of environmental investment and environmental policy-making (Source: National Working Conference of Environmental Protection Industry held on July 5, 2001 in Beijing).

[27] See details on the website of http://hzs.ndrc.gov.cn/hbcy/zhxx/t20050711_32473.htm.

combined economic and environmental improvements impossible and constitute market barriers and ineffective demands.

For the supply side attention should be given to the innovative potential of environmentally sound technologies. Generation, screening, and effective exploitation of ecoindustrial innovations depend on the underlying technological trajectories, the nature of market demands, and the characteristic of the organizations supplying them (Rennings, 2000). While technological investments are necessary, institutional and organizational innovation seem at the moment to be more crucial in helping to secure the direct and indirect benefits associated with the new technology, also by facilitating further incremental improvement along existing trajectories or paths. Successful innovations and their application depend on a combination of advances in scientific understanding, efficient cooperation between academic organizations and environmental industries, appropriate political and policy programs, smart strategy design, environment-oriented institutional changes and scale and direction of new investments.

In both supply and demand, policies and regulations can take a crucial position. Conflicts in the form of functional overlap and fragmentation among SETC/NDRC, SEPA and the local governments have existed for years, and we have elaborated upon them in terms of administrative ranking and conventional orbit. It is clear that China has not built up an effective and efficient administrative regulation system for EI's innovation and implementation, on either demand or supply side. Since no effective and efficient administrative coordination and policy harmonization take place among these organizationss, EI's innovative capacity has been undermined. There is a strong need for cleaning up the governance framework towards better performance. With the growing internationalization of the Chinese market for environmental goods and service, there is not much time to waste for the Chinese environmental industry.

REFERENCES

Alam, G. (2001). Acquisition of environmentally suitable technologies by China – the case of circulating fluidized bed combustion boilers, Report prepared for the United Nations Industrial Development Organization, UNIDO.

Asuka-Zhang, S. J. (1999). "Transfer of environmentally sound technologies from Japan to china," *Environmental Impact Assessment Review*, Vol.19, No.5, pp.553-567.

Chu, J. Y., Chen, J. and Zou, J. (2002). Future scale and market capacity of urban water environmental infrastructure in China: a system dynamic model. *Environmental Science*, vol.23, no. 4, pp.1-7.

Chen, J. and Liu, Y. (2001). "The barriers and opportunities of EI's innovation in China." In: *Proceedings of '21cn Forum - Green and Environmental Protection'*, The Chinese People's Political Consultative Conference (CPPCC), September 4-6, Beijing. (also published in China Machinery and Electronics Daily (Zhongguo jidian ribao), January 16, 2002, 6).

Chen, J. et al. (2001). Governmental regulation on commercialization of environmental infrastructure, *Research Report, Institute of Environmental Industry, Department of Environmental Science and Engineering*, Tsinghua University, China.

Chu, J. Y. and Chen, J. (2001). "Financing analysis for urban water environmental infrastructure," Environmental Protection (huanjing baohu), 290, December, pp.18-21. (also published in *Market Economy and Environmental Protection*, Chemical Industry Publisher, 2001, pp.158-162).

Diener, B. J. and Terkla, D. (2000). "*The environmental industry in Massachusetts: from rapid growth to maturity*", Corporate Environmental Strategy, Vol.7, No.3, pp.304-313.

ECOTEC (1999), The EU Eco-industry's Export Potential – a final report to DGXI of the European Commission, Birmingham: ECOTEC.

ECOTEC (2002), Analysis of the EU Eco-Industries, their Employment and Export Potential, Birmingham: ECOTEC.

Fu, T., Chen, J. and Zhang, L. Z. (2004). Misunderstanding of market principle adopted for urban water industry and government's role. *China Water and Wastewater*, vol.20, no.1, pp.88-91.

Fu, T., Chen, J. and Chang, M. (2005). Investment and financing mechanism in urban water industry and capital strategy. *Environmental Economy*, no.3, pp.31-35.

Guerin, T. F. (2001). "Transferring environmental technologies to China: recent developments and constraints," *Technological Forecasting and Social Change*, Vol.67, pp.55-75.

Gunningham, N. and Sinclair, D. (1998). "Designing environmental policy," in *Smart Regulation: Designing Environmental Policy,* edited by Gunningham, N., Grabosky, P. and Sinclair, D., Oxford: Clarendon Press, pp.375-453.

Ho, Peter and Eduard B. Vermeer (2006), *China's Limits to Growth: Greening State and Society*, Oxford: Blackwell Publishers.

Jahiel, A. R. (1998). "The organization of environmental protection in China," *The China Quarterly*, 156, December, pp.757-787.

Lieberthal, K. (1995). *Governing China: from Revolution through Reform.* W.-W. Norton and Company, Inc., New York.

Liu, Y. 2005. Phosphorus Flows in China: Physical Profiles and Environmental Regulation. PhD Thesis, *Environmental Policy Group, Department of Social Sciences*, Wageningen University, Wageningen, the Netherlands.

Ma, X. and Ortolano, L. (2000). Environmental Regulation in China: Institutions, Enforcement, and Compliance. Lanham [etc.]: Rowman and Littlefield Publishers, Inc.

Martens, S. (2006), "Public Participation with Chinese Characteristics: citizen-consumers in China's environmental management", *Environmental Politics* 15, 2, 212-231.

Martinsons, M. G., et al. (1997). "Hong Kong and China: emerging markets for environmental products and technologies", *Long Range Planning*, Vol.30, No.2, pp.277-290.

Mol, A.P.J. (2006), "Environment, modernity and transitional China. At the frontier of ecological modernization," *Development and Change.*, 37, 1, 29-56.

Mol, A.P.J. and D.A. Sonnenfeld (Eds) (2000), *Ecological Modernization Around the World: Perspectives and Critical Debates*, London: Frank Cass.

Mol, A.P.J. and Liu, Y. (2005). Institutionalizing cleaner production in China: the cleaner production promotion law. *International Journal of Environment and Sustainable Development,* vol.4, no.3, pp.227-245.

Murphy, J. and Gouldson, A. (2000). "Environmental policy and industrial innovation: integrating environmental and economy through ecological modernization," *Geoforum*, 31, pp.33-44.

Preston, J. T. (1997). “Technology innovation and environmental progress,” in *Thinking Ecologically - The Next Generation of Environmental Policy,* edited by Chertow, M. R. and Esty, D. C., Yale University, pp. 136-149.

Rennings, K. (2000). “Redefining innovation – eco-innovation research and the contribution from ecological economics,” *Ecological Economics*, 32, pp. 319-332.

SEPA (2001). The State of the Environmental Protection Relevant Industry Report in China. Beijing: SEPA.

SETC (2001). The Tenth Five Year Plan for the Development of Environmental Industry of China. Beijing: SETC.

Shenzhen EPB (2001). Investigative report on environmental protection relevant industries in Shenzhen city. Shenzhen: EPB.

Vermeer, E. B. (1998). “Industrial pollution in China and remedial policies,” *The China Quarterly,* 156, December, pp.952-985.

Wang, B. and Zhang T. Z. (2001). Valuing Dianchi lake of China with contingent valuation method, Research Report, Institute of System Analysis, Department of Environmental Science and Engineering, Tsinghua University, China.

Wang, Y. H., Yi, B., Li, B. J. and Teng, J. L. (2004). Development status of environmental protection industry in China. *Chinese Environmental Protection Industry*, no.10, pp. 36–38.

Zhang, K. M. (1992). *Report on China Environmental Investment*, Tsinghua University Publisher, Beijing.

In: Leapfrogging Development in Emerging Asia
Editor: Peter Ho, pp. 85-107
ISBN: 978-1-60456-410-5

Chapter 5

GREENING THROUGH INDUSTRIAL RELOCATION IN VIETNAM: THE CASE OF HO CHI MINH CITY

***Le Van Khoa*[1] *and Peter Ho*[2]**
[1]Environmental Protection Agency,
Ho Chi Minh City, Vietnam
[2]Centre for Development Studies and Faculty of Spatial Sciences
University of Groningen, The Netherlands

1. INTRODUCTION

Booming Industry, an Economic Blessing and an Environmental Burden

This contribution looks at the prospects of greening and leapfrogging development through the relocation of polluting industries in Vietnam. During the last two decades Vietnam has gone through substantial economic transformations. In 1986 the Sixth Congress of the Communist Party of Vietnam formally adopted the *Doi Moi*, thereby introducing an era of reform in order to end the enduring economic stagnation (Do 2002). The Doi Moi program's main focus was on market reforms, a gradual abandoning of the system of central planning and a controlled yet steady opening up of the Vietnamese economy to the outside world. Generally speaking Vietnam's economic reform policy can to date be considered a success, as Vietnam has managed to overcome the economic standstill and has moved into an accelerated process of industrialization and modernization.

Vietnam's industrial sector has made considerable contributions to the country's economic revival, which is reflected in a growth of the annual industrial output of approximately 13.56% in the period of 1996-2000. Even higher growth rates were achieved in the key economic areas of Vietnam, such as Ho Chi Minh City (Tran 2003; Statistical Office 2003). However, at the same time, Vietnam's rapid growth strategy of promoting industrialization and modernization is causing serious environmental problems. At an aggregated level Vietnam's industrial pollution might still be less severe than for example in its more industrialized neighbor Thailand; in the economic key areas environmental pollution is reaching alarming levels.

The situation is particularly problematic in Ho Chi Minh City (HCMC) where industry constitutes a major economic sector; accounting up to 49.1% of the city's GDP in the year 2003 (Statistical Office 2003). In addition to the most notable polluters in the heavy industrial sector such as iron and steelmills, pulp and paper factories, cement and chemical plants, the city's light industrial sector – involving food processing, electroplating, textile-manufacturing and dyeing, and leather and rubber processing – also contributes heavily to environmental pollution. A recent analysis of several local canals for example showed that the urban surface water does not meet the minimum Vietnamese standard, primarily due to pollution with untreated wastewater from domestic and industrial activities (DOSTE 2002). Moreover, air pollution from industry is increasingly acknowledged as a serious threat not only to the environment but also to public health. Vietnamese standards for total suspended particulate are exceeded in some areas of HCMC where the levels of CO are generally high as well (DOSTE 2002). In addition to mismanagement, an important reason for the rapid increase of industrial pollution is the limited application of (environmental) technology in production processes and waste treatment (Hung 1997). The capacity to invest in up-to-date cleaner technology is low, especially in small-sized enterprises, which constitute the majority of HCMC's industries.

In this respect it needs mentioning that HCMC's industrial landscape is characterized by a large number of industrial enterprises – 35,815 in 2003 – of which merely 935 can be considered as large-scale industries, the remainder being small- or medium-sized. To accommodate some of the industrial activity the local government of HCMC has constructed 12 industrial zones (IZs) with 2 export-processing zones (EPZs) where a total of approximately 800 enterprises are located (Hiep 2003). However, the major share of industries is located within residential areas in the 18 central urban districts of HCMC. The population density of these areas is high, with an average of 10,338 persons per square kilometer (www.pso.hochiminhcity.gov.vn). The fact that industrial activity is taking place in highly populated districts seriously adds to the severity of HCMC's pollution problem. Toxic contamination of air, soil, surface and groundwater and nuisances such as foul smells and noise, adversely affect the direct living environment and lead to high-risk situations for large groups of residents (Nguyen 1999). Moreover, the location of industrial activity in residential areas complicates the construction of infrastructure for cleaner production processes, both in terms of technology and management. It should not come as a surprise that the greening of industries by counteracting industrial pollution in residential areas has become a central concern for the HCMC authorities, and that the relocation of industries away from the inner districts is considered one of the priorites in environmental management.

2. Research Focus

Ho Chi Minh City's Industrial Relocation Program

In 1999, the People's Council of HCMC passed a resolution on "Environment and Health Protection" which stated that: "In order to reduce industrial wastes, it is necessary to speed up the construction of centralized industrial zones to enable the relocation of industry. Industrial waste must be treated in order to meet national environmental standards and all factories must introduce waste treatment by the year 2000" (DOSTE, UNIDO, and SIDA 1999). Various command and control instruments were implemented, such as regulations on the end-of-pipe treatment of waste and other emissions and forced closure of polluting enterprises.[1] In addition, some economic instruments, such as fines and energy-saving projects, and communication and support measures, such as the "Black Book" listing of notorious polluters, were introduced.

Since the environmental pressure from industrial activity seems to be especially problematic in densely populated areas, the focus of local authorities has been on tackling pollution problems in these districts. Since the late 1990s there have been several independent district-level programs for the relocation of polluting industries away from residential areas. Those programs in general arose as unplanned and incidental activities, initiated by the authorities of various districts in HCMC to immediately ameliorate severe industrial pollution cases. A case in point is Tan Binh district, which carried out its own relocation program between 1998 and 2001 under the authority of the District's People's Committee. Tan Binh has a population of 657,000, an area of 38.45 km^2, and about 4,500 enterprises within its borders and can be considered a relatively average district. By the end of 2001 a number of 387 enterprises had been ordered to shut down or to relocate away from the main residential areas. Although this was a significant achievement by the district authorities, the effects on environmental pollution appeared limited, and there was no decrease in the number of environmental complaints for residents. Partly, this can be explained by the rapid growth of industrial productivity in the district, which nullified the number of closed and relocated enterprises.[2]

Generally speaking the implementation of the *ad hoc* relocation programs that were developed at the district level moved slowly, and their results in improving the quality of the living environment have been limited. Explanations for the relatively small impact of these relocation programs are the lack of clear and integral environmental objectives, the absence of incentives for industries to participate in the relocation program and the fact that districts received no substantial commitment from the higher level of city authorities.

However, by early 2002 the People's Committee of HCMC expressed a strong interest in launching a new, more integrated relocation program for polluting industries, due to increasing environmental pressure and requirements for upgrading of the urban living environment. In a resolution dated March 14, 2002 the city's Party stated that:

> ... The standing committee of Party asserts that relocation of polluting enterprise is a right and logical policy, which has significant and important influence on both economic and social aspects. It is generally recognized that the city's People's Committee and industrial community should anticipate and overcome difficulties and barriers ahead.... This policy requires a virtual evolvement within 2-3 years, and is to be completed in 2004...
>
> (Saigon Giai Phong Newspaper, 2002)

On July 8, 2002, the city's People's Committee released decision No.80/2002/QĐ-UB on the approval of the 'Relocation of Polluting Enterprises to Industrial Zones Program'

[1] According to Nguyen Thien Nhan (1999) the authorities of HCMC shut down 115 industries in 1998.
[2] Based on interviews with Tan Binh District Authorities.

(hereafter called ROPETIZ program or relocation program). The program not only aims to effect an overall greening by solving environmental problems and promoting technological renewal in HCMC's industry, but also by combining the relocation of industries with changes in the city's economic structure and in the spatial organization of residential areas.

The concrete objectives of the relocation program can be summarized as follows:

i) to move all polluting enterprises, which are not able to engage in onsite improvement, to Industrial Zones and city perimeters by the end of 2004;
ii) to reorganize the planning, allocation and licensing for newly invested industrial projects, which are classified as polluting groups;
iii) to move small-scale handicraft enterprises to form handicraft villages and to create patterns of large-scale, stable operations, which are able to compete in the context of regional and global integration;
iv) to develop Industrial Zones in HCMC that adhere to the required planning. The integrated relocation plan applies to the whole city and includes various measures and activities, which will be discussed in more detail in a later section.

At present, the development and implementation of the relocation program are still ongoing, and any definitive conclusions on the success or failure of the program would therefore be premature. However, an in-between evaluation, as will be presented in this chapter, is valuable as it could point to the need for intermediate revisions of the program in order to improve its environmental impacts and to prevent unfavorable social-economic side effects. In this respect the objectives of the chapter in hand are: (i) to give an overview of the current state of implementation of the relocation program; (ii) to explain and analyze the institutional structure of the relocation program; (iii) to assess the opportunities and barriers for implementation of the program; and (iv) to come up with recommendations for the program's further development.

This chapter is based on empirical research that took place in the summer of 2003 (Lam, Huong, and Huy 2003). The empirical research included the investigation of 21 small- and medium-sized enterprises in two urban districts: District 11, and Tan Binh District. Moreover, interviews were conducted with officials of local- and district-level authorities and other environmental experts in the field. Various meetings between managers of industries and governmental officials were attended and observed. Empirical data are derived from a survey among 42 enterprises in three districts conducted by ENDA Third World (an international NGO for Environmental Development Action in Third Wold) in 2002.

The remainder of this chapter is divided into four sections. We will first describe the organizational setup of the ROPETIZ program in terms of the actors involved and their internal division of tasks and responsibilities. The basic introduction to the organization of the relocation program will help to understand the dynamics in the implementation of the relocation program, as described in the following section. This section provides an overview of the present state-of-affairs with respect to the actual relocation of industries in HCMC. It discusses occurring barriers in the various stages of the relocation program, from the perspectives of industries and from the authorities' point of view. The third section analyzes the institutional setting in which the ROPETIZ program is developed and implemented. The aim of this section is to see in which way the (in)formal task division between various governmental actors involved and their respective policy styles influence the implementation and success of the ROPETIZ program. In the final section, some conclusions and recommendations for the improvement of HCMC's relocation program are presented.

3. Organizational Outline of the ROPETIZ Program

Multiple Phases, Diverse Agencies, and Layered Responsibilities

In HCMC, the Environmental Management Division (EMD) of the Department of Science, Technology and Environment (DOSTE) is officially responsible for the management of environmental protection activities.[3] Its functions include appraisal and assessment of the environmental impacts of existing activities and new projects; the resolution of environmental disputes and problems caused by environmental incidents; pollution management, including the control of industrial discharges; training and community education on environmental protection; and monitoring of environmental quality (air, water, soil, noise). At the lower governmental level of city-districts the Environmental Group under the Urban Management Division (UMD) is involved in implementing the environmental plans and programs devised by DOSTE and/or the District's People Committees. The responsibility for HCMC's 14 Industrial Zones – which are spread throughout the city – remains with another authority, the HCMC's Export Processing and Industrial Zones Authority (HEPZA).[4] HEPZA is, with the support of DOSTE in charge of the environmental management of enterprises that are located in the special industrial and export processing zones.

Focusing specifically on HCMC's industrial relocation program it is clear that the institutional setup is relatively complicated. The ROPETIZ program is officially carried out by the People's Committee of HCMC and directed by a Steering Committee (chaired by the vice-chairman of the People's Committee with the deputy directors of the Department of Industry (DOI) and DOSTE acting as vice-chairpersons). Formally it is this Steering Committee that develops, monitors and supports the ROPETIZ program implementation. The committee holds the final authority to decide whether a certain industry needs to be relocated (or if onsite modernization will suffice), what the new location will be and which support the city will grant to the moving industry.[5] Under the leadership of the Steering Committee there are three divisions that have separate, yet sometimes overlapping responsibilities in the subsequent stages of the relocation program: the Basic Investigation Division, the Urban Planning Division and the Financial Policy Division. In addition, an Executive Board which belongs to DOI is established to help the Steering Committee in solving administrative activities of the relocation program. As is illustrated in figure 1, various governmental departments and organizations, whose tasks and activities will be described below, are involved in each of the three divisions.

[3] At present, DOSTE has become the Department of Natural Resources and Environment (DONRE).

[4] HEPZA is according to Decision No.731/Ttg dated October 3rd, 1996 established to carry out state administration on operations of export processing zones and industrial zones in HCMC territory in line with the principle of a "one stop service."

[5] One of the results of the Doi Moi reforms is that the large majority of HCMC's industries are at present privately owned. However, there are at the moment still 280 state-owned enterprises in HCMC, of which 128 are centrally state-owned. Relocation of these centrally state-owned industries cannot be decided by the Steering Committee, but should be discussed directly with the Ministry of Industry.

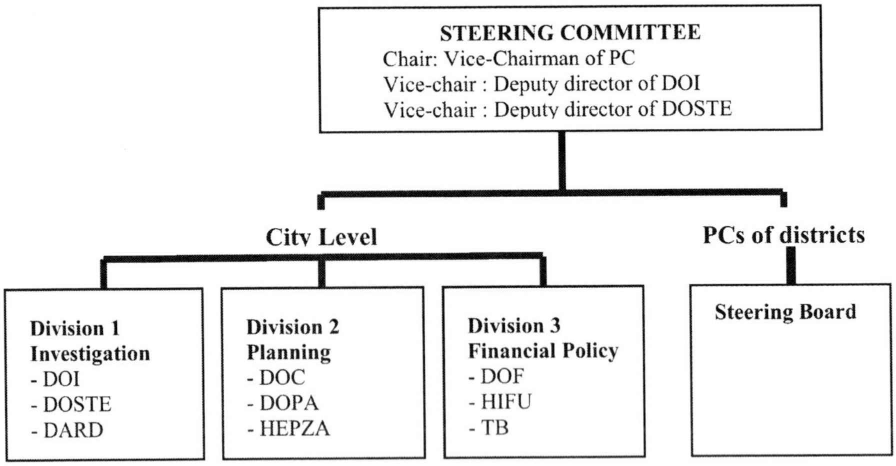

Source: People's Committee, 2002.

Figure 1. Organizational Structure of the ROPETIZ program.

Basic Investigation Division

In cooperation with DOSTE and PCs at district level DOI establishes the contacts with industries in HCMC and is responsible for collecting the initial data (such as basic industrial statistics) required for various assessments within the relocation program. Based on this information DOSTE checks the pollution levels of various industries to see whether onsite remedies are applicable or if industries have to be (partly) relocated. Decisions on the need to restructure, relocate, or even close down enterprises are made in line with the current standards on emissions and pollution levels, and with respect to the specific type of industry. Following DOSTE's recommendations experts of DOI (or the Department of Agricultural and Rural Development (DARD) for livestock breeding, forestry processing, agricultural and seafood processing) will make more detailed assessments of techniques, procedures and equipment needed for onsite improvement or relocation of specific industries. Subsequently, the Executive Board is responsible for synthesizing the comments of the various government departments, receiving and processing the feedback of enterprises, and submitting this to the Steering Committee.

Planning Division

The activities of the Planning Division are related to the spatial planning of the ROPETIZ program and the realization of the required 'hardware' in terms of infrastructure and the construction of industrial sites and zones. In this respect the Department of Planning and Architecture (DOPA) is responsible for the planning of Industrial Zones throughout HCMC and for redirecting the industries to these zones. At a more practical level the Department of Construction (DOC) is responsible for guiding enterprises through the legislative procedures for new construction and for solving obstacles related to the land-lease and land-use for industries in the relocation process. Once all procedures for the relocation have been settled,

HEPZA assists the industry in the actual resettlement in order to restart operations as soon as possible. Finally, it needs mentioning that the Department of Planning and Investment (DPI) controls the release of new business licenses and is thus in charge of keeping new unwanted industrial activities out of the urban areas.

Financial Policy Division

The tasks of the Policy Division are mostly related to the implementation of incentive and support policies as designed by the Department of Finance (DOF). In practice this largely boils down to the provision of financial advice and services to (potential) relocators. The HCMC Investment Fund for Urban Development (HIFU) will for example assist relocating industries in obtaining loans from credit organizations; experts of Tax Bureau (TB) will advise on preferential tax opportunities; and experts of the DOF can look into the applicability of various other financial support funds.

The above-described institutional arrangement is the formal structure for the implementation of the ROPETIZ program at the *city level*. However, in practice the district authorities play a crucial role in the concrete identification and relocation of polluting industries. Each city district is obliged to prepare a local relocation plan according to the city's master plan. The relocation plan should take into account the industrial pollution situation in the district and the measures that are already taken to reduce pollution levels. The city-level Steering Committee organized various meetings with districts' authorities to discuss the drafting and progress of local relocation programs, and also provided guidance in implementing relocation activities in each district. Usually the Economic Division and the Environmental Group (the latter is part of the Urban Management Division) of a district are jointly executing the programmatic activities, such as onsite visits, pollution measurements and decision-making. If the number of industries that are listed for relocation is particularly high in a certain district a specific management structure, mimicking the city-level structure (steering board) can be established. The district authorities mostly deal with small- and medium-sized industries, since most large-scale industries are internationally or state-owned and thus the responsibility of higher-level authorities.

4. Assessment of the ROPETIZ Program Implementation

In Between Progress and Stagnation

In this section the current-state-of-affairs and expected developments in the implementation of the relocation program will be discussed. The three stages of investigation and the planning and financial policy will be successively described and analyzed in terms of the progress made, as well as the difficulties and constraints encountered during the relocation process.

Identification and Investigation of Potential Relocating Industries

The Investigation Division performed the kickoff for the ROPETIZ program by a large-scale survey to identify major polluting industries in HCMC. Of 12,700 surveys sent out by DOSTE and the district PC's between February and May 2002, DOSTE received 7,164 replies from industries. Analysis of these data, supplemented with onsite investigations and discussion with local district authorities, indicated that there are about 3,000 industries in HCMC that do not meet pollution standards, of which about 1,200 have to be removed or relocated.[6] Two hundred sixty industries that are causing the most severe environmental pollution will be relocated in the first ROPETIZ period, between 2002 and 2004. The remaining enterprises have to move to industrial zones in a prospective second period of the relocation program, which has not yet been determined, or apply onsite measures to reduce pollution, such as end-of-pipe treatment and/or cleaner production technologies.

The decision to put industries up for relocation or closure is not only based on their failure to meet environmental standards, but also related to three other criteria regarding their location, the applied level of technology and the economic performance of an industry. More precisely the Ministry of Science, Technology and Environment (MOSTE) requires relocation, or closure for all industries that are: i) in a location which is not in line with the general urban planning at city and district level; ii) partly or completely operating with outdated technologies that cannot be improved through a feasible increase of investments; iii) not making sufficient profit to be considered effective in economic and social terms.

In addition to commanding individual industrial enterprises to relocate to industrial zones, the HCMC authorities also announced a list of 14 industrial sectors that are no longer allowed to operate in the inner districts. Enterprises in these sectors will be refused licenses for new business establishment in the residential areas of the city (see Box 1).

Box 1. Fourteen restricted categories of production in the inner city

1. Chemical industry (production of batteries, pesticides, paints etc.);
2. Waste recycling (paper, plastic and metal);
3. Fabric bleaching and dyeing;
4. Rubber vulcanization;
5. Leather tanning;
6. Electroplating and metal forgery;
7. Pulp production;
8. Production of building materials, pottery, porcelain and glass;
9. Wood processing;
10. Producing and processing of food products and beverages;
11. Tobacco processing;
12. Industrial breeding of livestock;
13. Animal slaughtering;
14. Coal processing.

Source: PC, 2002. Decision No.78/2002/QD-UB dated 08/06/2002 on announcement of the list of industrial sectors not be issued new investment license or certificate of trading.

[6] Updated calculation, August 2003.

This restriction on new industrial activity in the inner city implies that all presently existing enterprises from the banned sectors, are eventually obliged to follow the relocation plan of the city, or to shut down.

In August 2003, when the ROPETIZ program had been running for 19 months the Department of Industry published a progress report stating that 126 units had completed their relocation, 56 units had changed their produce to limit environmental impacts, and over 1,800 units had overcome their pollution problems by onsite improvements.[7]

Planning and Management of Industrial Zones

With regard to the planning and management of industrial zones, the other task of the ROPETIZ program carried out by the Planning Division, export processing zones (EPZs) and industrial zones (IZs) are of crucial importance, since these areas are designated to accommodate the industries that are voluntarily or compulsorily moving out of residential areas. The first EPZs and IZs in HCMC were established in 1991, and today there are two EPZs and 12 IZs with a total area of 4,000 ha spread over eight districts in HCMC. Currently, the authorities are trying to speed up the planning and construction of concentrated industrial zones, and of special handicraft villages for groups of small-scale enterprises. By July 2003 there were approximately 800 investment projects into these EPZs and IZs (foreign invested projects accounted for 46%) with a total investment fund of over VND 12,000 billion.[8] The occupancy rate of the industrial zones varies widely: 3 IZs are 100% filled, five IZs with 80% and four IZs serve only 50% of their total capacity (Hiep 2003).

In addition to various economic considerations the desire of the authorities to control and counteract industrial pollution was an important motivation in establishing the special industrial zones (Tran et. al. 2003). Frijns et al., (1997) remark in this respect that "the main rationale for relocation [to industrial zones] is that the concentration of polluting industries provides the opportunity for a common reaction to the challenge of environmental problems, by sharing costs and developing an efficient organizational structure for several small enterprises." Moreover, the concentration of industries entails some advantages for the authorities as well. The monitoring and management of industries in one central location is far more convenient than that of thousands of small firms scattered over a large area and in various districts.

Yet, environmental benefits are not the natural and automatic result of moving pollution industries to industrial zones. There are plenty examples to be found of cases where relocation programs merely resulted in increased pollution at a different place, duc to poor environmental facilities and lacking enforcement of environmental regulations in industrial zones (Bai 2002; Dasgupta 2000; Kwon and Lee 2003). To prevent such a situation the HCMC People's Committee issued a regulation on environmental management in IZs and EPZs. In this regulation (dated July 2002) it is stipulated that new industrial zones can only be put into operation after a detailed planning of the area has been completed. Moreover, facilities for water supply; wastewater drainage and –treatment; sites for the (temporary) storage of solid and toxic wastes; and personnel and material facilities to deal with

[7] According to the authors this figure of 1,800 should be reevaluated.

[8] USD 77.5 million, by the exchange rate of October 2003.

environmental calamities should be in place. Individual investors in Industrial and Export Processing Zones are obliged to complete waste treatment and storage facilities before commencing their activities. However, while the official procedures, standards, and guidelines for industrial zones are mostly in place, a major implementation deficit can be witnessed, leading to poor environmental improvements (Tran et. al. 2003). This is for example illustrated by the fact that today only two EPZs and two IZs operate central wastewater treatment plants.[9] According to the vice-director of HEPZA, Mr. Nguyen Cuong, the actual discharge of industrial wastewater at industrial zones is around 30,900 m^3 everyday, of which only 38.8 percent is treated (*Dau Tu* Magazine 2003).

One of the industrial zones that has constructed its own wastewater treatment plant is the Le Minh Xuan zone. This industrial zone was established in 1997 in Binh Chanh district and is operated by Binh Chanh Construction Investment Company that is 65% privately and 35% publicly owned. Presently there are 30 foreign and 84 domestic enterprises on the premises. The site's central wastewater treatment plant has a maximum capacity of 2,000 m^3 per day. Companies are obliged to pretreat their wastewater up to the discharge standards. The treatment costs for wastewater are VND 2,700 per m^3, while the price for water supplied to the companies is VND 3,200 per m^3.[10] The actual treatment costs charged to companies are based upon their initial water-use; calculated as 90 percent of the metered water supply. In practice there are some problems with this system, as a number of industries illegally exploits groundwater sources. Since the treatment charges are based on the *metered* water usage, such practices lead to financial losses for the industrial zones. Even more serious are the related environmental problems. Not only does the illegal use of groundwater result in low groundwater levels, defrauding companies discharge the excess industrial wastewater untreated into water bodies via their own sewers. As a result the water quality of canals in the vicinity of Le Minh Xuan IZ has continuously decreased over the years, changing from fresh water supplies into malodorous and toxic bodies of water. At present the IZ's management board seems unable to control the illegal discharges, due to a lack of capacity in the field of environmental management and monitoring.

Insufficient treatment of industrial wastewater is but one of the environmental problems caused by industrial zones. Other threats to the environment include air pollution and various problems related to solid waste. Factories in EPZ and IZs collectively generate about 63,000 tons of industrial solid waste each year, of which a substantial share is to be regarded as toxic wastes. At present procedures and facilities for the collection and treatment of these wastes leave a lot to be desired (*Dien Dan Doanh Nghiep* Magazine 2003).

Developers and managing authorities of industrial zones blame the delays in the construction of environmental facilities on the lack of financial resources.[11] Yet, although the high costs of environmental infrastructure are indeed a major constraint, there are also other explanations for the severe environmental situation in IZs and EPZs. According to some experts the developers and operators of industrial areas are simply not interested in environmental protection and do not accord any priority to investments in this field as

---

[9] Tan Thuan EPZ (with capacity of 10,000 m^3/day), Linh Trung EPZ (5,000 m^3/day), Le Minh Xuan IZ (2,000 m^3/day) and Tan Tao IZ (5,000 m^3/day).

[10] Respectively USD 0.206/m^3 and USD 0.174/m^3 (exchange rate, October 2003).

[11] According to the developer of Linh Trung EPZ, up to US 1 million has been spent on a wastewater treatment facility for Linh Trung 1. Le Minh Xuan IZ's developer has invested VND 11.8 billion (USD 760,000) to build a wastewater treatment plant (Saigon Times Daily 2003).

compared to, for example, expanding the capacity of their areas. The director of the Vietnamese Center of Environmental Technology and Management, Nguyen Trung Viet, stated that:

> Almost all IZs have been planned and are being operated without scant regard for the environment and with no or very little interest in its protection. Thereby many of them have destroyed the environment in the surrounding areas.
> (Dien dan Doanh Nghiep Magazine 2003)

The attitudes and opinions of the owners of (relocating) industries also contribute to the problem of pollution displacement. It appears that many managers of industry believe environmental expenditures (for end-of-pipe treatment, and cleaner production) are unproductive and unnecessary (Phung 2002). They often appear to have no plans for any (additional) environmental measures after relocating into industrial zones. Changes made in the organization of production processes are mostly aimed at increasing the productivity of existing technology. Generally speaking entrepreneurs consider relocation simply as an opportunity to expand their production and to invest (land, premises) in the future.

More and more it is becoming clear that while the city authorities are pushing the relocation of industries to IZs in the name of environmental protection, these industrial zones are in reality becoming disastrous "hot spots" of environmental pollution. Some government officials are by now publicly pointing out the shortcomings of the prestigious ROPETIZ program. A recent article in the Vietnam News quoted Doan Thi Toi, head of the Environmental Management section of DONRE:

> The scheme could mean a mere shift of pollution from urban areas to rural areas unless these businesses come up with solutions for their waste treatment.
> (Vietnam News, 2003)

A second, even more basic problem in the infrastructure planning of the ROPETIZ program is the limited capacity of industrial zones to accommodate the relocating businesses. According to some district officials, many small-scale industries face problems in finding new production sites (Labourer, *Nguoi Lao Dong* Newspaper 2002). Often suitable industrial zones are full and unable to accept new enterprises on the premises. Focusing for instance on the possibilities for the relocation of textile-dyeing industries, it becomes clear that only Hiep Phuoc IZ suits their particular production processes and environmental impacts. However at the moment Hiep Phuoc IZ lacks sufficient resources for water supply and has a very limited road-infrastructure. The construction of an 'economic-social industrial park' with 2,200 hectares of land in Cu Chi, Hoc Mon and Binh Chanh districts is under proposal, but so far only a needs-assessment has been completed. It is thus unlikely that the needs of the textile-dyeing businesses will be met during the duration of the ROPETIZ program (Lam, Huong and Huy 2003). The failure of many industries to find new production sites is also caused by the high land rental prices in IZs, that are at US$ 40-70 per m^2: out of reach for many small businesses.

Financial Setup of the Program: Support and Incentive Policies

The previous remark on land rental prices touches upon a crucial aspect of the ROPETIZ program, its financial structure and policies, which are during the implementation stage mostly the concern of the Support Policy division. The general financial design of the ROPETIZ program is based on the idea that government and industry have a shared responsibility in raising the required resources. Industries are expected to generate a share of the capital needed for the relocation and/or onsite improvements in environmental technology, for instance, through the sale of real estate and obsolete machinery. However, most of the relocating industries are small- and medium-sized enterprises (SMEs) that hardly own real estate or equipment, let alone property with a substantial market value (Saigon Times Daily Newspaper, 2002). Only the larger state-owned enterprises (SOEs) appear in the position to generate capital by selling their property and by transferring their land-use rights. In practice however, the procedures for selling SOE property are extremely complex, rendering it almost impossible to raise money in such a way (*Nguoi Lao dong* Newspaper 2002; Saigon Times Daily Newspaper 2002). The sale of SOE's land and facilities has to be proceeded through the Land Pricing Board, in order to fix a 'bottom price' for the land. Returns from the transfer of land-use rights (the value of the land) are to be kept in a temporary account at the State Treasury until the relevant authorities have approved reinvestment at a new location. Don Taylor (2002) remarked in relation to this that it is unclear what procedures apply to an enterprise that is 'asking permission' to utilize what, in essence, are its own resources.

As a result the financial backing of the relocation program is in reality almost solely the responsibility of the HCMC government that has developed four types of incentives and supports policies to (financially) encourage and facilitate the relocation of industries. First, the government withdraws each year approximately VND 200 billion (US$ 13 million) from its budget to compensate industries for the interest payments on loans contracted in relation to the ROPETIZ program.[12] Second, the authorities developed a series of tax benefits for enterprises that are participating in the relocation program. The business income tax is only 25%. The enterprise will be exempted from paying taxes for two years, and it will have to pay only 50% during another two years (four years for enterprises with more than 100 staff), and tax-free imports of equipment and technology not available in Vietnam. As a third incentive, businesses that are constructing new premises in industrial zones will be exempt from land rent for a period of three years, which can be extended to a six-year rent-free period for enterprises with more than 100 employees. And finally a special fund has been established to supply financial awards to industries that actively participate in the relocation program. This fund is not only designed to prize relocaters, but also to speed up the relocation process. According to their size, relocators could receive direct financial awards varying from VND 30 to 500 million (US$ 2,000 – 32,000) if they completed their relocation to a new site in 2002. For industries that complete their relocation process in 2003 or 2004, the awards are reduced to, respectively, 50% and 30% of the awards granted to similar sized industries in 2002.

[12] Loans used to build houses for workers will be subsidized with an annual interest of 3% for three years while loans for the construction of roads to new IZs, of sewage treatment plants and of public works will be free from interest rates for five years.

Although these support and incentive policies were already approved in July 2002, to date no concrete guidelines for their implementation have been issued (PC 2002).[13] It is unclear which government department is responsible for allocating the various funds and for handling the related administrative procedures. As a result relocating industries often remain unaware of support and incentive policies and are ignorant on the correct application procedures for (financial) assistance.

In the ENDA survey (2002) conducted among 42 enterprises in three districts a large majority of SMEs mentioned the lack of financial resources as the main obstacle to their relocation.[14] Outcomes of the research by Lam et al. (2003) among 21 SMEs in Tan Binh District and District 11 reinforce this conclusion (see Figure 2). When asked about the conditions, under which they would participate in the relocation scheme, the enterprises especially stressed the need for government support in terms of preferential loans, tax reduction, and an attractive price for leased land in IZs.[15]

However, of the 17 businesses (80.9%) that claimed to require loans only 11 enterprises had actually started the paperwork for application due to the complicated and unclear procedures (Lam, Huong and Huy 2003). According to HIFU, however, the main obstacles for SMEs to receive loans from formal sources are their unclear accounting papers, which make it difficult to properly appraise the business and the general insufficient provision of collateral security. Also city revolving funds have proven to date to be unattractive for enterprises, despite the zero percent interest. The major constraints of the use of these funds are the onerous credit appraisal procedures and requirements for collateral, which have discouraged enterprises from applying. The amount of monies on loan through these schemes are almost inevitably quite low in relation to the needs of the targeted (usually SOE) companies (Taylor 2002).

In addition to the high costs of the actual relocation process there are other, more long-term economic barriers for the relocation of industries. One of the strengths of many SMEs in HCMC is the flexibility of their product range and production process. Because they are located very close to the demand market they are able to adjust their production depending on the needs and immediate requirements of their customers.

Moreover, they can often buy their raw materials at competitive prices since there is a diversity of supplying sources in the vicinity. The often family-based SMEs have longstanding traditional and closed relations with customers and other firms at their current locations. If they move their enterprises to less centrally located industrial areas these traditional consumers and supplier sources may be jeopardized.[16] Other economic disadvantages of relocation include the fact that support services, such as machine repair and maintenance often remain more readily available within the city, and the increased transportation costs from the industrial areas to the owners' houses and markets (Lam, Huong

[13] Department of Finance-Pricing prepared a proposal of financial incentives to encourage and support relocators. The proposal was approved by the decision No.81/2002/QD-UB dated 08/7/2002 of People's Committee.

[14] See also interview with SME owners in Saigon Giai Phong News Paper, 7/2/2003.

[15] The other expectations for support concentrated on having a complete infrastructure in IZs, reasonable schedule to relocate, and conducting the ROPETIZ program at the same time to reduce unfair competition (Lam et al., 2003).

[16] This risk may be minimized by merely moving the production to industrial zones, while keeping storage and office space in the old neighborhood.

and Huy 2003; Frijns 2000).[17] Finally, many enterprises also face the possibility of losing skilled workers that can easily find jobs in similar industries closer to home. One third of the surveyed SMEs in Tan Binh District and District 11 believed it would prove problematic to hold on to their present personnel and to recruit new laborers after relocation. However at the same time about 60% of the interviewed SMEs assumed they could hire new laborers at the industrial zones against lower salaries (Lam, Huong and Huy 2003).

Support and Incentive	Mentioned by % of SMEs
Preferential loan	80.9
Tax reduction	71.4
Attractive price for leased land in IZs	71.4
Completed infrastructure in IZs	57.1
Reasonable schedule to relocate	57.1
Relocating at the same time	47.6
Fair enforcement of the Law	38.0
Other (larger leased land area, etc.)	19.0

Source: Lam et al., 2003.

Figure 2. Needs of 21 SMEs (Tan Binh district and District 11) for successful participation in the ROPETIZ program.

5. Policy Analysis

Key Actors, Shifting Responsibilities and Changing Policy Styles

In the previous sections the general goals and outlines of the ROPETIZ program have been described. Subsequently, we have identified various factors that influence the implementation of the relocation program. Based on this information, our evaluation of the relocation project is that some progress has been made, as illustrated by the number of businesses that have been relocated or have engaged in the greening of production processes. However, simultaneously the relocation targets set by the HCMC authorities had not been met at the time of writing. In this section we will assess to what extent the current configuration of policy actors involved in the development and implementation of the ROPETIZ program is hampering the achievements of the program.

Vietnamese policy networks have been described as structures that "can constantly be bypassed by party politics, informal networks, and unclear decision-making structures. This jeopardizes consistency, transparency, and reliability in policymaking and implementation, also in the field of environment" (Mol and Van Buuren 2003). Our focus will therefore not be on the formal institutional structure of the program, as was already discussed in section two, but instead on the concrete role-division between various government departments and levels. We will first look into the respective and interrelated tasks of three agencies that are most central to the ROPETIZ program: the People's Committee of HCMC, the Department of

[17] cf. the interview with Mrs. Nguyen Thi Ngoc Lan, owner of Ngoc Lan dyeing enterprise in Saigon Giai Phong News Paper, 7/2/2003.

Industry and the Department of Science, Technology and Environment. Secondly, the transfer of responsibilities from these city-level actors to more decentralized actors will be discussed. And the section will be concluded with some general observations regarding the changes in policy style that can be witnessed through the implementation of the ROPETIZ program.

Key-Actors: Internal Task-Divisions

In line with the formal organizational setup of the ROPETIZ program the People's Committee of HCMC can be considered a core actor in the actual program implementation. The commitment of the HCMC's PC is vital, as the committee employs its authority to pressure different governmental departments to actively cooperate in the relocation program and to mobilize resources for this purpose. The city PC's undisputed authority can often speed up and regulate the behavior of other actors involved in the program. Previously, the necessity of the involvement of HCMC's PC was already proven by the failure of independent district-level relocation programs (see section 1). District-level authorities lacked the backing of the city's PC and were thus unable to mobilize the required resources for successful relocation, such as the funds for incentive and support policies, and land for the construction of industrial sites.

However, the downside of the PC's steering power is formed by its traditional proclivity to the use of top-down, state-dominated approaches and instruments; leaving no room for flexibility and fine-tuning. SMEs – representing the large majority of industries involved in the ROPETIZ program – have specific needs in the relocation process, related to their limited capital and technology used, or their traditional relationship with customers and suppliers. Relocation of these smaller-sized industries will benefit from regulations that allow more room for individual variation and needs, than the generic regulations that are part and parcel of the PC's policy style. A case in point is the overall obligation for all concerned industries to relocate before the end of 2004. This common guideline has been proscribed without taking into account the availability of suitable alternatives for various sectors of industry. In the preceding section we have already seen how this posed problems to many textile-dyeing enterprises, as the construction of industrial zones that fit their requirements is still ongoing and not expected to be completed within the set time-limits.

At a more general level, the obvious haste of the People's Committee to complete the relocation process does not contribute to a smooth implementation of the ROPETIZ program. Not only is it impossible to prepare all 'hardware,' such as adequate industrial zones in time, the forced speed has also resulted in hampered soft-ware, in terms of underdeveloped support policies and unclear procedures within concerned government departments. DOI has for example been assigned a rather central position in the ROPETIZ program as the main executing agency. DOI is charged with the preparation of proposals for relocation (or onsite improvements) and for communicating these plans to the concerned industries. Thereby DOI seems to take on some tasks, which in terms of content, might be better left to the environmental experts of DOSTE. This central position of DOI seems to be more the result of informal politics, than of a logical consideration of various departmental capacities. On the one hand, this quaint task-division relieves the already overburdened DOSTE of some of its workload. On the other hand, however, we see that most of DOI's staff lack the required experience in the field of environmental management to adequately perform the duties assigned to them. In addition, considering the short timespan of the ROPETIZ program, the time to obtain relevant environmental knowledge and skills is extremely limited.

Although some of DOSTE's 'natural' tasks have been transferred to DOI, the environmental department remains responsible for the enforcement of environmental regulations, the checking of environmental licenses and the monitoring of pollution levels caused by potential relocators. These duties are mostly performed via a traditional command-and-control approach. However, recently DOSTE has also initiated more communicative and supportive approaches in, or related to the ROPETIZ program. One example is the promotion of green technologies for various sectors of industry, to enable enterprises to apply onsite improvements as an alternative for relocation. Taking into account the outcomes of the ENDA survey (2002), which pointed out that that 70% of the questioned SMEs is unwilling to relocate and would instead prefer onsite improvements to counteract environmental pollution, this can be considered an important new step within the ROPETIZ framework.

In addition to its regulatory enforcement mechanisms, since 1994 DOSTE has also published Black Books of polluting enterprises in HCMC. By publishing the names of the most notorious polluters DOSTE aims to employ public exposure as an incentive for greening industries. However, a survey conducted with the 87 firms that were listed in the 1997 Black Book gives reason to question the results of the Black Book strategy [21]. The survey revealed that 90% of the factories believed that a listing in the Black Book did not affect their business. And 84% of the enterprises stated that their listing has had no negative effects on loan negotiations with financial institutions. Even more remarkable is the fact that 39% of the factories thinks it may just as well be an advantage to be listed in the Black Books, as this lends weight to claims on governmental subsidies to implement pollution-reduction measures.

Generally speaking we may conclude that DOSTE is striving to take on more proactive and flexible roles in the ROPETIZ program rather than merely enforcing environmental legislation. However, so far the actual results of additional innovative approaches have been limited. Due to severe understaffing, financial constraints and poor management DOSTE's activities are in practice more or less confined to reacting to requests from the HCMC-PC or to citizen complaints.

Decentralization: Shifting Roles and Responsibilities

Although the ROPETIZ program is still largely implemented in a traditional, centralized policy fashion, with major responsibilities and powers vested in city-level authorities, other governance styles are gradually appearing in HCMC's environmental policy. As mentioned previously, HCMC has recently issued new environmental regulations for industrial zones. Traditionally the responsibility for the implementation of these regulations would have been delegated to the city-level Department of Science Technology and Environment. However, the city government has now made a few tentative steps towards a decentralization process by transferring these duties to the HCMC Export Processing and Industrial Zones Management Authorities (HEPZA).

Many researchers in development studies consider the decentralization of policymaking and –implementation as a positive development, despite the fact that decentralization is not without problems (Litvack 1998). With respect to the decentralized management of HCMC's industrial zones, Frijns (2003) remarked that to date a unified system of environmental legislation is lacking, which has resulted in a large variety of environmental management between zones. Moreover, HEPZA appears ill-equipped to control and orchestrate the environmental governance and performance of the 14 industrial and export processing zones, not in the least because the expansion of their duties has remained unaccompanied by a substantial increase in their budget (Taylor 2002).

HEPZA is not solely accountable for the bad environmental record in the majority of industrial zones. Although, the responsibility for the implementation of environmental regulations in industrial zones is delegated to HEPZA, DOSTE remains officially in charge of the enforcement of environmental legislation. In this respect, the official policy is to regularly inspect and control the operation of the enterprises. Both the Industrial Zone Management Board and DOSTE have to submit their monitoring reports to the People's Committee. In practice, however, deviations from the official policy and procedures seem to be the rule rather than the exception. There are known cases of heavily polluting industrial zones that repeatedly fail to comply with the obligations, without ever having to face the consequences (Tran et al. 2003).

Decentralization of duties and powers in the ROPETIZ program is also effectuated through the involvement of district authorities. Districts' People's Committees (Districts' PC) are responsible to cooperate with the Steering Committee to develop local relocation plans in order to enable and promote the relocation of polluting SMEs in the district. In this respect they can be regarded as key actors that decide which enterprises have to move, to implement onsite improvement, or even to shut down their entire operation. The Districts' PC's attitude towards relocation is crucial, and if it is seriously interested in the implementation, it can create a suitable and flexible approach to accelerate and facilitate the ROPETIZ program. A case in point is Tan Binh District, where the district authorities employed a community-based approach for the relocation of several textile-dyeing enterprises. The district authorities including the Division of Economy, the Division of Urban Management and the Management Board of Tan Binh IZ, arranged in cooperation with DOSTE many open discussions with SME owners to find out SME's specific needs in the relocation process. Through this open, participatory approach industries were given the chance to voice their opinions and needs, while the authorities received valuable feedback to effectively adjust their policies (Lam, Huong and Huy 2003).

The outcome of the discussion between the Tan Binh authorities and the textile-dyeing industries is that the existing Tan Binh IZ will be expanded to exclusively accommodate this specific group of relocaters. A central wastewater treatment plant will be established, jointly operated and paid for by representatives of the Management Board of Tan Binh IZ and the group of textile-dyeing SMEs. A service fee will be paid by the enterprises. Through this approach, the dyeing SMEs are guaranteed a stable operation at a new location. A further advantage of this approach is that it regenerates inner city areas and allows the people involved in small- and family-scale enterprises to remain within the geographical vicinity of their residence (Taylor 2002).

A final novelty in the ROPETIZ program is the involvement of industrial branch associations in the development and facilitation of relocation activities. The most striking example in this respect is the proactive stance and conduct of HCMC Plastic Association (HPA). Plastic enterprises are facing various difficulties in the relocation process, such as excessive land rental prices, insufficient infrastructure and complex administrative procedures in industrial zones. Instead of awaiting directions and support from the city authorities, the sector-association decided to act upon itself and bring together 13 plastic producers to construct a new production site in Long An Province. The project aims to meet the relocation and expansion requirements of several plastic producers.[18] The initiators claim that the industry-focused "cluster zone," known as the Duc Hoa Ha Complex, will be cheaper and more practical than renting in industrial zones (IZs).

We wanted to extend our existing plant in Binh Chanh, but this location is already crammed, and any growth would eat away our space for storage of raw materials and finished products," said Thuong Chi Minh, director of plastics packaging company A Chau. "But we couldn't afford the high rents charged at most IZs. So when the HPA asked us to invest in their plastics complex, so we signed up for 24,000 m^2."

The HPA say a dozen other plastic producers, mostly privately owned, have also signed up, and the 54 ha site is almost full. There seem to be two main reasons for plastic producers to favor the Duc Hoa Ha Complex: the provision of cheap land-use rights and the zone's simple, flexible administrative procedures. HPA will see to all paperwork and red tape, freeing up investors to get on with their business. HPA general secretary Tran Cong Hoang Quoc Trang states that since the tenants are also the joint complex owners, the complex will be built quickly and looked after carefully. According to Trang the association has now received approval from the Long An People's Committee to expand the complex to 130 ha (Vietnam News, 2002).

Changing State-Industry Relations: A New Policy Style?

HCMC's authorities and governmental departments appear genuinely committed to achieving the ROPETIZ targets. However, while willing and ambitious bureaucrats and politicians have been designing policies and instruments in the framework of the ROPETIZ program, they have failed to simultaneously secure the co-operation of industries. The framework of the relocation project is largely based on the traditional top-down design and insufficiently tuned to the needs of Ho Chi Minh City's (small- and medium-sized) industries. Both in the development and implementation of the ROPETIZ program the relationship between city authorities and industries has been loose and mostly one-sided. The absence of constructive interactions between authorities and industries that allow feedback on policies, instruments and support systems in the ROPETIZ program is a serious predicament to the project implementation.

However, at the same time signs of a transforming policy style can be witnessed in the implementation of the relocation-program, such as the decentralization of environmental management in industrial zones, the community-based approach in Tan Binh district, and the emergence of new roles for sector associations. These are examples of innovative, flexible strategies of negotiation and consultation that complement the otherwise command-and-control approach of the ROPETIZ program and give new shape to the relationship between government and private sector. Such policy innovations might herald a comprehensive multidimensional process of political modernization, implying new relationships between state-industry and state nongovernmental actors with more decentralized, flexible and consensus-oriented arrangements. Yet, at the moment less ad hoc and institutionalized

[18] According to an article in Vietnam News (27/3/2002).

channels for public participation in decision-making processes are still lacking in the relocation program, as they are in HCMC's environmental management in general. And recounting on universal changes in the government's policy style would thus be premature.

CONCLUSION

In the greening of industries and improving urban environmental quality, a great number of rapidly developing cities have relocated polluting industries to suburban districts or surrounding smaller cities. This strategy also includes the relocation of industrial wastes to central treatment facilities that are located outside of city boundaries (Bai 2002; Kwon and Lee 2003; Darliana 2003; Morikiho 2003). Similar to the closure of industries, relocation can prevent pollution exposure and reduce high-risk situations for local residents. In addition to improving the environmental quality in residential areas, clustering of industries in special zones can facilitate government's monitoring and control of industrial pollution.

There can also be several economic benefits attached to the reorganization of a city's industrial landscape. If managed properly, the relocation can be a good chance for industries with obsolete facilities to update production lines and install treatment capacities that discharge less pollution. The results of the survey of SMEs in District 11 and Tan Binh District showed that almost all entrepreneurs expect that relocation will be an opportunity to enlarge their production scale and to increase their production area (Lam, Huong and Huy 2003). The ROPETIZ program aims furthermore to revitalize suburban economies, as the relocation of industry – and with that employment – will increase the income of citizens in HCMC's suburbs. Finally, industrial relocation can be an aspect of local development and land-use planning policies when industries that occupy a large area in city-centers become the target of land use redevelopment (Darliana 2003). This is the case with VISSAN Company in HCMC. This large slaughterhouse occupies a great plot of land in the city's tourist center, which could be more profitably developed, for instance, as an amusement center.

Yet, in the preceding sections various risks and problems related to relocation in general and Ho Chi Minh City's ROPETIZ program in specific have also been identified. One of the most important reservations in this respect relates to the environmental benefits that can be expected from the ROPETIZ program. Previous research has pointed out that if the environmental monitoring and management in the designated industrial areas is lacking, industrial relocation will merely result in the relocation of pollution (Dasgupta 2000; Frijns and van Vliet 1999). This risk is a harsh reality in HCMC, where the environmental management of industrial zones leaves a lot to be desired. A case in point is the experience with poor wastewater treatment in Le Minh Xuan Industrial Zone, as described in this chapter.

With regard to the potential economic benefits of the relocation program the picture is not entirely favorable either. Relocation usually imposes relatively high costs to industries – especially to SMEs that are generally short in capital – and is therefore likely to meet with considerable resistance (Berkel 1996). Also in HCMC small-scale industries generally claim to be unable to generate their share of the funds needed for relocation (Lam, Huong and Huy 2003; ENDA 2002). Moreover, the majority of industries appear to consider responses to environmental challenges as an additional cost and inhibition to their competitiveness, rather than perceiving it as an opportunity to revive and renew their business strategy. It is thus not an established fact that industries, big or small, will invest in cleaner production processes after relocation.

Generally speaking, the city government has expressed strong commitment to the improvement of local environmental quality by launching a citywide industrial relocation program. However, it has been overambitious by setting an overly strict timeframe for the execution of the ROPETIZ program in the period of 2002-2004. As a familiar proverb says, 'haste makes waste.' The relocation project suffers from incomplete preparation in terms of the realization of infrastructure, the construction of relocation sites, the establishment of support mechanisms, and sufficient commitment from businesses. Instead of trying to restructure the entire city at once, it would have been more feasible to first select a few districts that have "hot spot" pollution problems as demonstration sites for the relocation program. Such an approach would have avoided a fragmentation of scarce human and financial resources and would have enabled a more gradual buildup of experiences and capacities.

However, in order to enhance the outcomes of the integrated ROPETIZ–program in its current setup, consultations with sector institutions, branch representatives and industries would help in drafting a clear step-by-step schedule for further project implementation. A more interactive approach could aid to pinpoint bottlenecks and come up with reasonable incentives, and the necessary support to speed up the current relocation process. At present the ROPETIZ program is almost solely designed according to a regulatory, command-and-control approach, with only limited backing of market- and information-based strategies. By building in various feedback mechanisms the mix of instruments can be better tuned to the needs and characteristics of different types of industries.

However, it should not only be the industrial sector that determines which instruments are to be employed, but also the present capacities of environmental authorities and other sectoral departments. If the HCMC Investment Fund for Urban Development (HIFU) is not equipped to assist relocating industries in obtaining loans due to limited human resources or lacking insight into the needs of industries, this should be taken into account in credit-application procedures. Simultaneously, the involved governmental agencies have to work to build institutional capacities, such as their capabilities in the field of policy- and decision-making, and to enhance their management skills, in terms of monitoring and enforcement of legislation.

Our final comment on the ROPETIZ program is without doubt most elementary and concerns the choice of relocation as a major strategy in environmental management. The authorities of HCMC have chosen to emphasize the relocation of polluters as their main instrument in the battle for improved environmental quality (of residential areas). It might be more effective, both in environmental and economic terms, to focus on the promotion of green technologies as a strategy to counteract industrial pollution as also described in the chapter by Liu et al. elsewhere in this volume. Although there are numerous barriers to the adoption of green technologies there is also a long list of benefits (Frijns et al. 1997; Blackman 2000). Green technologies do not only help to decrease the output of waste per unit input, they also prevent the loss and inefficient use of materials. Moreover, it improves the working conditions for laborers (Berkel 1996). Interviews with industrial entrepreneurs have pointed out that they often prefer the installment of green technologies to the relocation of their enterprise. Yet, simultaneously industrial managers lack knowledge on the subject, and fear the high cost that they expect to come along with green technologies (see also section 4) (ENDA 2002; Lam, Huong and Hoy 2003). According to Phung (2002) the promotion of green technologies at the company level in Vietnam is hampered by a lack of specific policies to encourage these practices, a lack of market incentives and information, and limited access to technology. Efforts from local- and state-level authorities to even out these barriers would unquestionably benefit a long-term solution to HCMC's pollution problems.

REFERENCES

Bai, X. (2002, January 6). Industrial relocation in Asia. *Environment*, p. 8.

Berkel, C.W.M.V. (1996), *Cleaner production in practice: Methodology Development for Environmental Improvement of Industrial Production and Evaluation of Practical Experiences*. Amsterdam: University of Amsterdam Press.

Blackman, A. (2000), Informal Sector Pollution Control: What Policy Options Do We Have? *World Development,* 28, 12, 2067-2082.

Darliana, E. (2003), Industrial Relocation Effort to Minimize Environmental Pollution – Case study: NGAGEL Industrial Area Surabaya, Indonesia. *Kitakyushu Initiative Seminar on Industrial Relocation.* HCMC, Vietnam.

Dasgupta, N. (2000), Environmental Enforcement and Small Industries in India: Reworking the Problem in the Poverty Context. *World Development, 28, 5*, 945-967.

Dasgupta S. *et al.* (2000), *Small Plants, Industrial Pollution and Poverty*. Hillary R. (ed.), Small and Medium-Sized Enterprises and the Environment: Business Imperatives, Greenleaf Publishing.

Dau Tu Magazine, 4 August, 2003.

Dien Dan Doanh Nghiep Magazine, No.39/2003.

Do, et al. (2002) The Doi Moi Process and Human Development, *Vietnam's Socio-Economic Development*, no. 29, spring 2002.

DOSTE (2002), *Annual Monitoring Report*, HCMC.

DOSTE, UNIDO and SIDA (1999), *Cleaner Production Case Studies, Food Processing, Pulp and Paper, and Textile Processing Sectors, Reduction of Industrial Pollution in HCMC,* Project TF/VIE/97/001, HCMC.

ENDA (2002), *Public Consultation Report: Financial Mechanism and Incentives for Supporting the Polluting Industries Relocation.* UNDP-Project VIE/96/023 Environmental Management in HCMC, HCMC.

Frijns, J. et al. (1997), *The Institutional Environment of Small-Scale Industries*. Frijns J. and J.M. Malombe (eds.), Cleaner Production and Small Enterprise Development in Kenya, Nairobi, pp. 49-59.

Frijns, J. and B. van Vliet (1999), *Small-Scale Industry and Cleaner Production Strategies,* World Development, Vol. 27, No. 6, pp. 967-983.

Frijns, J. (2000), *Pollution Control of Small-scale Industry in HCMC: To Relocate or to Renovate,* Paper International Conference on Industry and Environment in Vietnam, HCMC.

Frijns, J. (2003) Relocation or Renovation: Greening Small- and Medium-Sized Enterprises. Mol, A.P.J. and J.C.L. van Buuren (eds.), *Greening Industrialization in Asian Transitional economies. China and Vietnam*, Oxford: Lexington Books.

Hiep, V.T. (2003), *HCMC Industrial Zones Development and Relocation and Resettlement Situation.* Presentation in 4th Thematic Seminar: Kitakyushu Initiative Seminar on Industrial Relocation. HCMC, Vietnam Aug. 2003.

Hung, N.T. (1997), *Industrial Development and Natural Environmental Protection, Environmental Auditing*, "Workshop on Mitigation of Industrial Pollution," Economic Development Institute Worldbank.

Kwon, C. and S.H. Lee (2003), *The Experiences of Industrial Relocation in Korean Cities, with Special Reference on ULSAN Metropolitan City,* Kitakyushu Initiative Seminar on Industrial Relocation. HCMC, Vietnam Aug. 2003.

Labourer (*Nguoi Lao Dong*) Newspaper, July 25, 2002.

Lam, N.T., T.T.L. Huong and N.Q. Huy (2003), *The Implementation of the ROPETIZ relocation Program for SMEs in HCMC, Vietnam.* Case Studies. Supervised Internship Evaluation. Nong Lam University –Wageningen University.

Litvack J., et al. (1998), *Rethinking Decentralization in Developing Countries,* World Bank, Washington D.C.

Ministry of Planning and Investment (MPI) (2001), *Public Disclosure of Environmental Performance in Development Planning: Lessons from HCMC's Black and Green Book. Technical Report No.3,* Hanoi.

Mitchell, C. (2003), *Promoting Cleaner Production in Vietnam: The Role of Training and Education in Strengthening Industry's Environmental Behaviour,* University of Toronto.

Mol, A.P.J. and J.C.L. van Buuren (2003), *Greening Industrialization in Asian Transitional Economies,* Oxford: Lexington Books.

Morikiho, I. (2003), *Relocation of Industries to Kanazawa Industrial Park,* Kitakyushu Initiative Seminar on Industrial Relocation. HCMC, Vietnam, Aug. 2003.

Nguoi Lao dong Newspaper, 26 December 2002.

Nguyen, K.K. (1999), *Status of the System of Vietnam's Environmental Standards: Obstacles in its Application and Future Directions,* paper presented at the first Asean Environmental Forum, September 20-24, Hanoi.

PC, 2002.

Phung Thuy Phuong (2002), *Ecological Modernisation of Industrial Estates in Vietnam,* Wageningen University.

www.pso.hochiminhcity.gov.vn

Saigon Giai Phong Newspaper, 2 April 2002.

---, 7 February 2003.

Saigon Times Daily Newspaper, 8 April 2002.

---, 2 June 2003.

Statistical Office (2003), *Statistical Yearbook.* HCMC.

Taylor, D. (2002), *Financial Mechanisms for the Relocation of Heavily Polluting Industries,* Project VIE/96/023, UNDP.

Tran, T.M.D. et al. (2003), *Environmental Management for Industrial Zones in Vietnam,* In: Mol A.P.J. and J.C.L. van Buuren (eds.), *Greening Industrialization in Asian Transitional Economies,* Oxford: Lexington Books.

Tran, T.M.D. (2003), *Greening Food Processing Industry in Vietnam: Putting Industrial Ecology to Work,* Wageningen University.

Vietnam News, 12/8/2003.

---, 27/3/2002.

In: Leapfrogging Development in Emerging Asia
Editor: Peter Ho, pp. 109-128
ISBN: 978-1-60456-410-5

Chapter 6

Environmental Governance in the Information Technology Sector: The Case of Hsinchu, A Science-Based Industrial Park in Taiwan

Wen-Ling Tu
Department of Public Policy and Management
Shih-Hsin University, Taiwan

1. Introduction

The information technology (IT) industry has been viewed as the world's most dominant industrial sector with its intensive capital and technology advantages. The globalization of the IT industry, along with its actual and potential environmental impacts, has rapidly influenced the economies of many developing or newly industrial regions, such as India, China, Singapore, Malaysia, and Taiwan. While many have discussed and celebrated the success of IT industrial development, this chapter explores the dark side of the industry. Using a case study approach, it examines high-tech development and its environmental consequences in Taiwan. The chapter argues that characteristics of IT development – its network-based specialized production, regional competitiveness, and quickly changing nature, which create unique economic advantages – are posing great challenges to the greening of industries, let alone, the leapfrogging of development. In Taiwan, these challenges are further amplified: the IT sector is nurtured by a political and economic scheme, which adds to its dominant power.

This chapter will first review the industrial characteristics of IT and its environmental implication. It will then give a brief overview of the success of Hsinchu Science-Based Industrial Park (HSIP) in Taiwan and discuss how the formation of IT's industrial dominance has changed social and political dynamics. Environmental consequences resulting from high-tech development are identified in the following sections. My argument stresses that the challenges to environmental planning and governance are rooted in the IT-dominant structure according to which the state is in favor of IT promotions and the capacity to govern the environment is consequently weakened. In addition, there has been a lack of regional environmental capacity control in response to rapid high-tech expansion; while active public participation in environmental issues is not encouraged for fear that this might impede the rapid tempo of development. The research focuses on what follows high-tech economic success, as well as the environmental constraints associated with it. It gives a glimpse of pivotal environmental processes in high-tech dominant contexts, and may contribute to a more sophisticated and detailed understanding of the pitfalls of environmental planning and management in a high-tech political economic context.

2. Literature Review

Characteristics of the IT Industry and Their Environmental Implication

The global IT economy is one of the major forces transforming the world social order. It includes the phenomenal growth of the Internet, large volumes of cross-border information flows, and the wireless revolution. The dynamic nature of the IT industry delivers substantial economic benefits not only to IT manufacturers and software companies that design products based on computers, but also to general users of those computers. In many developing countries, the IT industry has provided the opportunity for national industrial upgrading and gives global competitive advantages; it has become the backbone of the industrialization process. As Jussawalla (2003) describes, the influential power of IT industry drives the processes cut across all sectors and impacts virtually all human and societal activities. In addition, the industry also reshapes the international division of labor, generates new growth patterns, and in the process, spawns new products, jobs, and livelihoods.

2.1. Industrial Characteristics of Information Technology

Speed

Industrial IT development is different from other manufacturing industries in terms of its speed and scope. The speed at which new products and technologies are introduced is the key that drives fierce competition for market share. The IT firms work hard to gain market share and earn monopoly-like profits by being "first movers," that is releasing products several months ahead of competitors. Moore's Law, which decrees smaller and faster chips, indicates that the pace of IT product innovation would double about every eighteen months. Speed matters, as it relates to the industry's unique innovative characteristics. Speed also matters for firms to be able to compete and survive. Time pressure accelerates product cycles. Rapid technological innovation and just-in-time manufacturing shorten the production cycle as well as the life-span of IT products. Brief production cycles and fierce competition in a highly uncertain market impel a unique mode of industrial organization that is structured by a network of flexible specialization.

Flexible specialization, illustrated by industry analysts, best characterizes the uniqueness of the high-tech industrial structure (Mazurek 1999; Saxenian 1996; Luthje 2002). Because the product cycles are too short and technology changes too quickly for large integrated firms to respond effectively, flexible firms based on a decentralized network of manufacturing are best suited to such market swings. Given the demand for just-in-time manufacturing and rapid product turnover, flexibility is an imperative in the IT industry. Manufacturing takes less centralized forms than integrated mass production under the same roof of a plant. Flexible specialization aims not only at the flexibility to switch product lines in response to changing demand, but also at the assurance of quality of output with adaptive process capability for specialized IT products. Economic geographers contend that flexible specialization creates tremendous economic advantages for driving continuous innovation and production.

Scope

The scope of IT development is reflected in its cluster phenomenon, broad industrial interrelations, and increasingly global expansion. It has been widely observed by regional planning theorists that the spatial arrangement of IT development often follows the structure of an agglomeration or cluster (Saxenian 1996; Hall 1998; Wheeler *et al.* 2000). Regional planning theorists use transactions cost analysis to explain the incentives for high-tech firms to agglomerate, because proximity creates positive externalities by reducing some direct costs and some less tangible costs. The region cultivates industry growth by creating an available set of inputs including a skilled workforce, institutions, and waste disposal facilities that further attract investment and development in the region. A successful IT industrial cluster also generates upstream linkages to local suppliers, driving regional industrial development.

The IT sector, as defined by the standard industrial codes manual, includes hardware, software, and communications (CAP 2002, 90). Interdependence among software firms and computer manufacturers spans the production of semiconductors, disk drives, circuit boards, and video display equipment. The development of IT products has also spawned thousands of businesses in related industrial sectors, such as the manufacture of chemicals and materials used in IT production.

Unprecedented globalization has seen an acceleration in the size and scale of the IT industry. The industry's flexible specialization drives global expansion in order to lower labor costs, diversify risks, or come closer to markets. Its widespread reliance on strategic alliance for production, marketing, and even waste treatment, in the form of outsourcing and subcontracting, creates long and complex supply webs that span many countries around the world. Today IT components are being manufactured, assembled, and used around the world. While the higher-tech and most research-intensive facilities are located in the USA, Japan, and some countries in Western Europe, IT manufacturing has expanded to many countries throughout Europe, Asia, Central and South America in the hierarchical tiers of the IT industry (Mazurek 1999).[1]

2.2. The Environmental Implication of IT Industrial Characteristics

Economic Advantage, Environmental Disadvantage

While economists and regional planners celebrate the great mobility of the capital, the flexible organization of specialization networks, speedy development, and the cluster structure that contribute to economic advantage in the global economic restructuring process,

[1] Countries include, for example, Ireland, Italy, Spain, the Czech Republic, Hungary, China, Malaysia, Korea, Singapore, Taiwan, Thailand, Israel, Costa Rica, Mexico, Brazil, etc.

economic advantages may become environmental disadvantages when planning and policy initiatives fail to address the environmental challenges of these new industrial characteristics. Emerging environmental constraints can only be understood through a broader analysis of industrial characteristics and environmental challenges associated with the industry.

Speed both drives rapid growth for the IT industry, but also poses new challenges for planners and policy makers. On the one hand, the industry makes intensive use of chemical compounds, many of which are unknown quantities in terms of their potential risks to the environment and health. Mazurek, in her analysis of microchip production, maintains that there is no preset formula for making successive generations of microchips (Mazurek 1999, 27). Each new chip generation is achieved through continual experimentation and refinement, or 'learning.' Roughly one-third of all new microchips require chemistry and equipment that are completely different from those used to make their predecessors. While production processes change continually, risk assessment takes years of trials. On the other hand, in a rapidly changing industry where profit margins are razor-thin, the firms are unwilling to divulge internal data for fear of that information being passed on to competitors. This in turn hinders scientific understanding of the risks to community and environmental health. These factors combined mean that policy makers and planners have been far behind in identifying the potential problems of IT production and have failed to assess and identify risks in time.

In addition, accelerated product cycles and rapid changes in technology have resulted in short life-spans for IT products, which creates tremendous waste problems. An estimated 12.75 million computers were recycled in 2002, and most of the old computers were exported to developing countries such as China, India, and Pakistan for disassembly, according to the report, "Exporting Harm: The High Tech Trashing of Asia" (BAN 2002). As computers contain substances such as lead, cadmium, chromium, mercury, and plastics, it is difficult to dispose of them safely in a landfill or by incineration. This creates tremendous environmental justice concerns because many developing countries lack the capacity and regulations (or, usually, the political will) to implement sound environmental practices.

The issue is one of increasing concern due to the rapid global expansion and cluster nature of the industry. Success in high-tech development often contributes to rapid expansion that requires more resource support from the region. While clustering fosters idea sharing and social network building that further benefit continuous tech innovation and rapid industrial growth (Saxenian 1991; 1996), the overall environmental capacity, and water and energy supplies to support the expansion of cluster are often not taken into consideration. To satisfy the needs of the IT industry and gain regional advantage, regions often make promises and offers beyond their capacities and means.

Flexible specialization drives tiered outsourcing and subcontracting, creating difficulties in tracing environmental responsibilities, measuring performance, and preventing problems. Subcontractors tend to be small or medium-sized firms, with thin profit margins and less capacity to dedicate to monitoring environmental and social concerns. Cooperation through information and technology sharing in streamline production for best environmental practice is hindered due to the quick-changing and trade-secretive nature of the industry. Driven as it is by trade initiatives and high-tech company expansion, the globalized scope of the electronics economy and its great mobility of capital allow firms to escape environmental and labor restrictions through choice of locality. Global IT operating standards for community, labor, and environmental health protection often counter the downward pressure of international competition. The IT industry's great transnational mobility further grants more bargaining power to the industry in negotiation with governments. As governments aim to promote the competitiveness of the industry and earmark their resources for IT use, there is a general lack of environmental perspective in terms of addressing the negative impacts of IT development. As the IT industry continues to revolutionize societies across political borders, it becomes more difficult to adequately regulate and manage its environmental and social problems.

IT Success Overshadows Environmental Disadvantages

The industry's environmental disadvantages are obviously surpassed by the appearance of economic success; it has the ability to create tremendous wealth and job opportunities. The glitzy image of IT development is further reinforced by its clean and knowledge-based outward look. Unlike traditional industry, associated with images of smoke and undesirable factory buildings, IT industrial parks are often located in picturesque surroundings with green open spaces and park-like landscaping. They are often described as being "smokestack free," with an RandD emphasis and high profit margins, as well as many jobs that are highly paid and highly skilled, especially relative to other market regions in many developing areas.

With all this anticipation for IT development, the development of IT high-tech parks becomes an opportunity for leapfrogging, particularly as many governments are placing greater emphasis on the industry to propel their regional economies (Jussawalla 2003). High-tech IT parks, featuring technological research, investment, and production, are aimed at bringing together all dynamic network externalities for incremental development. The "high-tech recipe," originally from Silicon Valley, which "combines such ingredients as a research university, a science park, and venture capital in an environment free of government regulation or labor unions" (Saxenian 1991, 38) has been followed globally in many regions in the form of IT parks to promote regional economy and development. The development vision to create high-tech centers such as Silicon Island in Taiwan, Silicon Glen in Scotland, and Silicon Mesa in Mexico reflects the desirability of the "silicon" blueprint in terms of the wealth and prosperity rapid and extensive IT development brings about.

The erection and expansion of high-tech parks, with their aforementioned economic importance and social image, have often remained unchallenged. In such a political-economic context, viable alternatives can hardly be based on public concerns about industrial competitiveness and national security. Rational planning cannot be implemented because of the powerful influence IT has on public policy, which often overrides the environmental and social concerns addressed in the planning and policy making processes. Although IT development in many regions was originally initiated and developed by governmental agencies with various kinds of subsidies and support, the desire for more rapid economic development has further propelled the industry into the global production network (Evans 1997; Gopalan 2000; Wangel 2001; Foran 2001). The powerful image and operation of the industry in social, economic, and political domains has further led to a gravitational pull of resources that has reshaped the social and political dynamics between the industry, governmental agencies, and communities.

The study of HSIP demonstrates that the downsides of high-tech development are hardly addressed on the policy level. Complex environmental issues associated with broad social patterns of production, consumption, and habitation generally fall outside the purview of environmental planning and management.

3. HSIP Success and IT Industrial Dominance

Hsinchu Science-Based Industrial Park of Taiwan is frequently cited as a "miracle" of the information technology era (Saxenian 1999). HSIP, the first science-based industrial park in Taiwan, has been known for its success in nurturing high-tech industries since it opened on December 15, 1980. The overall performance of HSIP in the past two decades has been impressive. In 2001, the most dismal year on record for global info-tech industries, the enterprises in the park contributed to a total revenue of US$19.6 billion, representing an annual decline of 34 percent. In 2004, the total revenue of 384 park companies reached US$ 32.5 billion, representing a growth rate of 27 percent.[2]

As the birthplace of Taiwan high-tech, HSIP has propelled the country into the competitive position of global high-tech landscape. To date, Taiwan high-tech companies produce the majority of the world's notebook computers, motherboards and monitors, and a range of other electronic-related products. According to figures released by the Institute for Information Industry Market Intelligence (MIC) in 2005, Taiwan has achieved astoundingly high global market shares in a number of major IT product categories, including motherboards (98.3%), notebook computer (82.4%), world's WLAN cards (89%), internal VOIP phone routers (83%), etc.[3] The high-tech industry has been a strong contributor to domestic economics and the transformation of Taiwan's industrial structure. It is viewed as the driving force behind the country's economic growth.

The success of the Hsinchu model has been widely discussed. The dominant accounts in the literature focus on active government policy, such as support from the Industrial Technology and Research Institute (ITRI), a publicly funded RandD organization (Mathews 1997), or public subsidy, such as tax incentives and loans (Chen 1990). Scholars also attribute successes in IT-related industries to a concentrated, highly skilled labor supply, a network of ancillary companies, and a potential market (Castells and Hall 1994). Moreover, the social and professional networks of an international technical community, advice from overseas Chinese engineers, and the advantage of cluster development are commonly recognized factors in Taiwan's industrial IT success (Saxenian 2001). In recent years, Taiwan's interdependence with the Silicon Valley economy and its particular split type of industrial structure have been increasingly discussed (Saxenian 1999). For example, Saxenian argues that the adaptive capacity of Taiwan's IT sector derives from the decentralization of the industrial system and its close economic connections with Silicon Valley (Saxenian 2001).

The literature reveals the fact that, given the nature of volatile, uncertain, globalized high-tech development, Taiwan has succeeded in gaining a comparatively advantageous position in the international division of labor in the high-tech economy. The state and the high-tech sector have been the two key players in creating Taiwan's high-tech landscape. The state's influence stems from its well-designed planning and policy promotion, whereas the high-tech sector's impact results from its growing network and global economic expansion.

---

[2] Data available at eweb.sipa.gov.tw/en/investment/about/development.jsp, accessed on 12 January 2005.

[3] Data available at investintaiwan.nat.gov.tw/en/news/200511/2005110901.html (official site of Ministry of Economic Affairs). Accessed on 12 January 2006.

State-Industrial Coalition for IT Promotion

The emergence of the dominant industrial IT structure in Taiwan has been shaped by the state's unparalleled IT promotion policies and internationalized IT growth and competitiveness. The Science-Based Industrial Park (SIP) Policy provides IT firms with the best investment environment in Taiwan, featuring tax breaks, one-stop services, and cheap rents for high-tech firms.[4] The high-tech park is a special district directly managed by the central authority that leverages the park as it enjoys more preferential treatment. The privilege of the HSIP is presented in the spatial form of well-planned road systems, public facilities, low-density housing, large-scale open spaces, and its own school system.[5] On the planning and policy level, the government further promotes expansion of the industry with assurance of resources for IT use, to pursue the national policy objective of transforming Taiwan into a "Green Silicon Island."

The IT promotion policies boost the importance of the IT sector's role in reshaping Taiwan's economics. My research in HSIP indicates that the success of HSIP creates high-tech illusions that further contribute to the IT dominance structure. In such a context, the IT industrial processes shape the mobility of people and capital, drive business–political coalescence, and dominate the allocation of resources. The industrial sector possesses tremendous power and resources in social, economic, and epistemological areas, which prompt central and local authorities to gravitate toward it. It is especially powerful when the industry is embedded in a global network of production that results in great flexibility of capital movement. The globalization of capital has minimized the role of the nation-state in the decision-making realm. In the case of Taiwan, the state plays a comparatively small part in controlling and directing the development of high-tech industry because it fears capital flight. To encourage the firms to stay and keep the industry vigorous, government has little option but to continue or expand its favorable treatment of high-tech firms.

IT dominance poses unprecedented challenges to an already weak system of environmental governance in a developmental state such as Taiwan. As discussed in the following section, policy circles have suffered from a lack of environmental perspective addressing the negative impacts of the industrial IT development. Public interests in environmental, community, and occupational health issues are often left out of the coalition policy in the IT-dominant context. With many jobs at stake, many IT workers who are also community members keep silent about their concerns on health issues, while IT's socially glitzy image and fear of capital flight further reinforce the trend. With its power over

[4] Industries in general enjoy the "Encouraging Investment Statute" and the "Statute for Promoting Industrial Upgrade." In additional to these two statutes, firms in the park also benefit from the IT Duty-Free Law and Land Rental Regulation (Kao 1995 and Chien 1997). The IT Duty-Free Law gives every new firm investing in SIP a five-year tax break followed by a maximum tax rate of 22 percent. After its first-time investment, the company can receive an extra four-year tax break for its additional investment. The whole production line from importing raw materials, production machines, to exporting final products, all benefits from tax break policies. From 1990 to 1994, HSIP's firms paid 1.57 percent of their sales profits to the Ministry of Finance compared to 15.29 per cent on average by the top 100 manufacturing industries, and 20 percent by the small business sector (Hsia 2000). The Land Rental Regulation provides cheap rental for high-tech firms with about US$1.27 sq. feet per year (Chang et al 2001).

[5] The National Experimental High School, established in 1983, provides education for the children of employees working in the Park and nearby research institutes. The school is divided into a high school, a junior high school, a primary school, and a kindergarten, as well as a bilingual section (Chang 2000).

resources and sophisticated organization, IT's hegemonic power presents unprecedented constraints for environmental mobilization to challenge and change its imperative dominance.

4. Environmental Impacts of High-Tech Development

4.1. Toxic Release

Hundreds of chemicals, including several highly toxic substances, are used daily by HSIP firms; it is no wonder that HSIP has been suspected as the primary source of pollution in the Hsinchu region.[6] According to Ku, inadequate treatment of wastewater discharge by HSIP has resulted in substantial land and air pollution (Ku 2002). Wastewater generated by HSIP firms includes heavy metal wastewater, organic solvent wastewater, and acid/ alkaline waste liquid. Vaporous releases of volatile organic compounds (VOCs) and emissions of other toxic gases from high-tech production process and wastewater discharge are major sources of air pollution in the Hsinchu region.

Pollution problems were not publicly revealed until 1997, when wastewater from HSIP was found illegally discharged into a local irrigation ditch and creek. Moreover, a fire at United Microelectronics Corp. (UMC) in HSIP also shocked the Hsinchu residents with its toxic gas release.[7] Local communities began to question the once widely held myth of the high-tech industry as a "clean" form of economic development. While most information about levels of pollutant emission and their impact is not available, several environmental catastrophes, such as pungent smells from the Kerya River, unnatural sex changes in Ke-Yin conches, numerous fire incidents, abnormal blood test results in community members, and repetitive dead fish incidents in Kerya River after 1997, suggest environmental degradation.

Pollution problems are generated from high-tech production process. Various types of pollution commingle, which then result in broad, adverse impacts on the region's environment and its community's health. Several local papers have reported that residents shut windows and close doors in order to resist the pungent odors that result in nausea; sometimes they seek treatments from local doctors.[8] Individual complaints have been filed in community forums generally not attended by high-tech firms. Local and national government agencies have found environmental management and regulation extremely challenging in the face of the rapid growth of high-tech production.

The environmental catastrophes reflect the insufficiency of current environmental control systems in the face of rapid development in the high-tech sector. Wastewater discharge by HSIP increased fourfold between 1997 and 2000.[9] According to Science-Based Industrial

---

[6] According to the Environmental Bureau of Hsinchu City Government (2000), major pollutants emitted from HSIP include isopropyl alcohol, acetone, 2-butanone, n-butyl acetate, nitric acid, ammonia, hexafluroethane, etc.

[7] The UMC fire took place at the end of July 1997. An interview with a firefighter revealed the serious pollution problem of toxic release as he explained, "A week after the fire, I walked into the factory without wearing a mask. I fainted and was sent to the emergency room and hospitalized for a week," (Chang *et al.* 2001).

[8] For example, news headline was shown as "Park's Wastewater release Pungent Odor, KaoFong Li's Residents Hold Breath, Some Residents Nausea and Seek Treatment." (*China Times,* 1998/11/11).

[9] Effluent from HSIP was 105,960 cubic meters daily (cmd) in 2001, while the original capacity of the wastewater treatment plant was 18,000 cmd (Liu *et al.* 2001; SIPA 1997).

Park Administration (SIPA), water flow in the Kerya River (upstream of HSIP) is 106,000 cubic meters daily (cmd), but HSIP alone would discharge 185,000 cmd wastewater that could become a serious environmental burden for the Kerya River (SIPA 2002). SIPA has tried to expand and improve wastewater treatment facilities in response to the rapid growth of IT production.[10] However, wastewater treatment plants can only treat some chemicals and monitor suspended solids (S.S), chemical oxygen demand (COD), and biochemical oxygen demand (BOD), without examining volatile organic compounds (VOC), which are the major negative byproducts of IC (integrated circuit) production.[11] As an analyst in the Industrial Technology and Research Institute (ITRI) pointed out, "IC production uses many toxic chemicals, and because of patent right, even the importers do not know what ingredients are inside. The use of chemical keeps changing (and) that results in complex environmental consequences. Even discharged wastewater is qualified in terms of environmental standards; it may still be toxic," (Chiu 2000, 115).

The pollution incidents also showcased defects of environmental governing capacity in HSIP as well as in environmental systems in Taiwan. SIPA's reputation in handling pollution control was damaged when half (ten out of twenty) biochemical treatment plants were discovered to be out of order with seriously reduced water filtering function in the water treatment plants (Chang et al. 2001). According to investigation by Chang *et al.* (2001, 20), water flow monitoring systems from wastewater treatment plants have been broken ever since installation. The Shengli incident further implicated the consequence of mismanagement of hazardous waste.[12] According to the cabinet's Research, Development, and Evaluation Commission, the loopholes of toxic waste management include low capacity for waste treatment, which has led to widespread illegal dumping of waste; the inability of local agencies to monitor firms' waste management thus enabling companies to skirt rules; and inadequacy of monitoring processes licensed treatment firms often commissions illegal entities to dump waste haphazardly (SIPA 2002).

In the environmental controversies, SIPA and related responsible agencies were acting as defenders for the HSIP firms, seeking solutions by improving public facilities for pollution treatment instead of requiring companies prevent pollution. In response to Shengli incident, SIPA started to build an incinerator inside the HSIP to treat industrial sludge and waste solvent in December 2001. It further raised controversies locally because eight schools of varying levels are within 2 km radius of the incinerator and local people were not informed until late 2004.[13] It is evident that, in terms of environmental management, SIPA is a service

[10] According to the HSIP Environmental Impact Analysis and Response Report, the second phase in the expansion of the wastewater treatment plant was completed by 1998 to handle 110,000 cmd. The third phase of the expansion project increases treatment capacity up to 55,000 cmd, and another wastewater treatment plant is being built in the third phase of the development of HSIP that handles 20,000 cmd. At total 185,000 cmd capacity will meet the long-term need for wastewater treatment (SIPA 2002).

[11] According to EPA, industrial wastewater discharge was examined in relation to thirty-six substances. The HSIP's wastewater treatment plant only handles COD, BOD, and S.S, while control of the other thirty-three substances should be carried out by each individual plant. Among these thirty-six substances, there is no control on VOC and no guarantee that there is no VOC in the water (Chiu 2000, 114).

[12] An illegal toxic dumping caused Kaoshiung city to be without any usable water supply for two days. As Shengli had signed contracts with 80 percent of IT companies in HSIP to deal with high-tech firms' waste solvents, all waste solvent (1,500 metric tons per month) was temporarily stored in HSIP after the incident and shipped to sludge farms in South Taiwan. SIPA then planned to build an incinerator for drying sludge and reusing the waste solvent as fuel (*United Daily* 2000/7/21).

[13] Local map of eight schools and the incinerator is posted on Hsinchu Foundation's website http://www2.hsinchu.org.tw/f-image/location.jpg.

provider rather than a manager for companies in HSIP. Related public agencies also tried to accommodate the needs of the high-tech sector, particularly when the firms complained that the tightening of environmental regulations would raise costs, forcing them to seek more affordable investments outside of Taiwan. No politician who has ever confronted HSIP has succeeded in seeking election or reelection at the local level.[14] On the issue of wastewater discharge, no individual company was ever held responsible for the catastrophic events, while SIPA bore all the blame, leaving the firms unscathed.

Although a series of environmental catastrophes and public health concerns in Hsinchu has awakened local residents from the myth of a green, clean high-tech development, the problems have not been made known nationwide. Media coverage shows a bias toward HSIP that tends to describe its brighter side. Lin, in his analysis of news reports on HSIP, found that environmental issues only occupied an average of 15 percent of the total HSIP news coverage from 1991 to March, 2002. Reports on two-thirds of incidents involving HSIP environmental disputes never make the national headlines. Rather, these stories appear only in local Hsinchu newspapers (Lin 2002, 106). In fact, during the Shengli incident, news reports made no effort to investigate solvent treatment at HSIP.[15] Instead, the high-tech companies in HSIP were legitimized as victims of the incident rather than waste solvent suppliers. On the issue of coastal pollution, the news reports focus on how the research report caused huge losses to oyster farmers. There were no inquiries on the nature of the pollution or who the polluters are. When political figures ate raw oysters to show support for oyster farmers, the research was discredited for exaggerating health hazards associated with coastal pollution.

The image of "high-tech equals low pollution" is still widely accepted by the public. Unbalanced media coverage reflects not only this social tolerance toward the rapid-growing IC industries, but also the influences of the social-politically powerful high-tech sector. Some environmental activists have felt extremely frustrated that high-tech pollution problems were downplayed or ignored by the national press. Even after several catastrophic events had occurred, the high-tech firms still claimed that companies in HSIP were the environmental role models of "high profit with zero pollution."[16] The powerful influence of the high-tech sector was particularly evident when the illegal discharge of wastewater was revealed in 1997. A reporter from the *China Times* wrote a series of articles to question water pollution from HSIP. *China Times* was then boycotted and could not receive any advertisement from companies affiliated with HSIP for a month. As a result, the high-level managers of *China Times* visited HSIP in person and apologized to the HSIP firms' representatives. The reporter was forced to change his news route to avoid continually uncovering negative impacts of the HSIP (Lin 2002).

---

[14] The city government's tension with HSIP is generally believed to be one of the causes of the defeat of Mayor Tsai when seeking re-election. In addition, Mr. Wu and Mr. Chen, who revealed HSIP wastewater discharge records, failed when running for election as local aldermen.

[15] In 2000/7/22 Coolloud (a social movement internet media) column "HSIP turned itself from 'waste producer' to victim of Shengli incident," it states that "when the HSIP got impaired, all media and all governmental officials rushed to seek solutions for HSIP's waste solvent. They seemed to dismiss common sense. They didn't even ask where those waste solvents that HSIP companies contract to Shengli went …" The article can be accessed at http://www.coolloud.org.tw/news/trackback.asp?tbID=440.

[16] The CEO of UMC wrote a public letter of accusation to the Hsinchu city government. The letter was advertised on all major newspapers in May 15th 2000, stating that "companies in the HSIP are all proenvironment and that to develop zero pollution, high-profit industries (IT production) not only contribute to economic growth, but also help the environment recover," (Chen 2002, 129).

4.2. Land Use

All resource allocation proposed by local and national governments has promoted high-tech development, as the IC sector is viewed as the "locomotive of economic transformation." Cluster land development, driven by the economic success of HSIP, has been at the forefront of local strategies to promote local growth. Such development has dramatically changed the landscape from agricultural use to the industrial zone in Hsinchu area. The capital has rapidly expanded as the industry is increasingly embedded in the global production network. The land has been rapidly developed in response to global market demand, and the state has provided land and public infrastructure to satisfy the needs of industries. National land use planning and other policies are encouraging industrial development in the name of high-tech cluster for regional competitiveness. In periods of economic downturn, the nation provides more subsidies and economic incentives to stimulate development projects. "Fighting for economic development," promoting high-tech in particular, has been the most important task for the Democratic Progressive Party (DPP) government in winning its ruling legitimacy and recognition.[17] In a newborn democratic nation such as Taiwan, economic growth is the main driver of all developmental policies, often at the expense of social reforms.

It is clear that site selection for the industry is not being based on sustainable land use or regional environmental objectives. The environmental agencies have been powerless to stop inappropriate land development in water protection areas or hillside land. In the Quanta Environmental Impact Assessment (EIA) incident,[18] the President stated that higher local EIA standards would hinder business development.[19] The incident showcased the power of the high-tech sector related to its commitment to investment, and how environmental laws can be set aside to accommodate the demands of business. Like the UMC EIA incident, violation of the EIA law has never become the major concern in the public domain or in policy-making processes. Instead, how to mitigate the loss of companies as a result of the implementation of EIA was the focus issue. The EIA has taken a back seat as the state has relaxed land use controls (e.g., on hillside land use), released national lands for large areas of high-tech park development (e.g., Taiwan sugar lands), simplified the permit application process, and lowered land rents to encourage industrial investment, echoing the policy principle of "keeping industry rooted in Taiwan."

4.3. Water Planning and Allocation

The issue of water shortage has recurred every winter and spring since 1995. The problems of water shortage have led to complaints from HSIP firms, which spent a hundred million New Taiwan Dollar (NTD) on trucking water in (Chen 2001). The situation was getting worse in 2002. In spring, the Ministry of Economic Affairs (MOEA) implemented emergency measures in response to the drought to transport water from farms to HSIP.

---

[17] The DPP became the ruling party in 2000 after KMT's fifty-year one-party rule. It was the first political power transition at the national level in Taiwan.

[18] Quanta is one of the major thin-film transit-liquid crystal display (TFT-LCD) producers in Taiwan.

[19] The President stated that this stone (which was slowing down the EIA process) should be removed and that he could get down on his knees and beg those local government officials for those business, if necessary (*Taipei Times* 2001/8/17).

Around 15,000 hectares of Hsinchu farmland were left fallow, but high-tech firms such as IC manufacturers were exempted from restricted measures.[20] At the end of 2002, the Cabinet ordered 3,000 hectares of farmland to be left fallow in order to reserve water supplies for industrial and residential uses (Taipei Times 2002/12/12). MOEA further decided to leave 28,000 hectares of agricultural land in the north unseeded, at a total cost of 1.5 billion NTD in compensation for affected farmers, to ensure enough water for the economically important HSIP. In August of 2004, nightmare of water cuts occurred again. This time, muddy water caused by torrential rains paralyzed all of the water company's purification systems, severely hitting northern high-tech area and leaving people without water usage for a week. The high-tech firms were forced to spend large amount of money to buy water in order to maintain their production lines (Taipei Times 2004/9/3).

Industrial expansion has demanded more water and electricity that has resulted in further resource scarcity. Worsening water fights in Hsinchu are indicative of a rapid growth in high-tech industries that require increasing resource investment to meet the demands of rapid global market expansion. Seventy to eighty percent of water is mainly supplied for IC manufacturing in HSIP, as it requires ample ultrapure water for its manufacturing processes (Chen 2002, 82). Since 1994, a number of 8 inch wafer manufacturing plants have started operation. Investment in 12 inch wafer manufacturing plants became popular after 1998. Water demand increased 140 percent between 1996 and 2000 (Chen 2001). According to the CEO of UMC, water demand for HSIP would increase to 2,230,000 cmd and would require 700 million cubic meters of water a year to satisfy the demand of HSIP.[21]

In the face of water shortages, the firms complained that unstable water and electricity supplies have impeded their investment in Taiwan. They blamed the government's inability to provide production elements, comparing it to its competitor states such as China and Singapore.[22] Fearing that drought would hit high-tech production and result in a further loss of "international economic competitiveness" and "economic prosperity," the government took two measures to ensure the water supply for the high-tech sectors.

First, it shifted water from agricultural use to HSIP use. According to the Water Resource Act, agricultural water supplies should be prioritized over industrial supplies (Article 18 and Article 20) when there is a conflict for water usage.[23] However, the policy was altered when high-tech came into consideration. According to the Council for Economic Planning and Development, the agricultural sector uses around 70 to 80 percent of the nation's water but accounts for less than 3 percent of GDP. There would not be much concern about cutting its water supply (*Taipei Times* 2002/5/3). On the other hand, high production value industries such as IC and thin-film transit-liquid crystal display (TFT-LCD) were exempt from the restriction in order to guarantee Taiwan's international competitiveness (*Taipei Times* 2002/5/3). The policy echoed the claims of the IT sector that high-tech should be given first priority for resources because of its higher production value.

---

[20] In February 2002, Deputy Prime Minister announce fallow plan to solve water shortage problem of HSIP (*Industrial and Business* 2002/02/28).

[21] Domestic water usage for 1.3 million people in the Hsinchu and Maioli areas is 413,300 tons per day (source from replying message dated April 17, 1998, Water Resource Bureau of Ministry of Economic Affairs, in response to an inquiry about water supply and demand in the Hsinchu and Maioli areas).

[22] Complaints are often viewed on newspapers when water shortage occurred (e.g. *China Times* 2002/2/28).

[23] Under the Water Resource Act, the order for water users' rights is: 1. residential and public water supply; 2. agricultural water supply; 3. hydrological purpose; 4. industrial water supply; 5. transportation purpose and other functions.

Second, to accommodate the rapid expansion of the high-tech industry, the government keeps investing in water resource infrastructure. In 1994, the government proposed to HSIP short-, mid-, and long-term plans to expand water supply projects, at an estimated cost of 17.2 billion NTD. In 2002, proposed plans for increasing water supplies include overland water shipping, fallow project, dam construction, and desalinator construction[24] to ensure ample water supplies to HSIP (Chen 2002, 62). The increasing development of the water supply infrastructure raised environmental concerns in the communities as three existing and proposed dams are located in Baoshan village, the neighboring village to HSIP.

The state, again, is the service provider for the IT firms as they are the most economic important sector for the country's competitiveness. Water supply controversies show the favoritism given to the high-tech industry.[25] The high-tech sector has been given first priority for scarce resources and the support of public infrastructure, policies, and opinion. The state used its authority to solve the HSIP water crisis. Besides farmers' protests and requests to increase compensation, little effort has been made to challenge the unjust water allocation and environmental impacts. Environmental and social interests are marginalized while economic interests dominate national development.

5. Challenges to Environmental Governance

The issues of environmental and resource use controversies associated with HSIP challenge its glitzy images of "low pollution and low energy consumption." They also reflect ineffective environmental governance in relation to HSIP. This section analyzes and summarizes the problems of the state's environmental systems in the face of HSIP. I argue that the challenges of environmental governance are rooted in IT-dominant structures that result in the setting aside of environmental regulation, planning, and implementation. The problems are examined in the political-economic context of Taiwan to illustrate the social forces that have crippled environmental governance.

5.1. Setting Aside Environmental Standards in Face of the High-Tech Development

In the face of public inquiries on environmental catastrophes, the HSIP's high-tech firms and SIPA kept claiming, "everything is in accordance with laws." However, "complying with the laws" does not guarantee a safe environment; rather, it means a lack of standards to regulate certain pollutants. In terms of water pollution control, the laws only regulate S.S, COD, BOD, acids/alkalines, and thirty-three other heavy metals. Many chemical compounds were not included in the examining list (Chiu 2000). In the case of air pollution control, major points of air quality monitoring include SO_2, NO_2, CO, and O_3, which are different from

[24] The high-tech firms dislike the project because the water cost would be higher from desalinator. By 2005, the government is still in negotiation with the high-tech firms for the "acceptable" price (*United Daily* 2005/10/19).

[25] As Vice Premier Lin said, "The government would come up with a plan to put rice fields out of commission in favor of keeping computer chips and other products rolling off production lines at the science park ..."(*Industrial and Business* 2002/02/28).

pollutants emitted by HSIP. As Ku observes, there is no comprehensive planning for chemical use management, largely because the regulatory control is often far behind industrial application, and the health impacts of the toxic substances are not well known. In the face of the short production cycles and intensive toxic use that characterize high-tech development, the environmental system has failed to develop comprehensive toxin inventories for monitoring and controlling toxic release (Ku 2002).

Laws and regulations are the baseline for the companies to follow. However, firms often complain that tightening of environmental regulations would raise costs, forcing them to seek more affordable investments outside of Taiwan. Law making and policy implementation have been thoroughly interfered with by the industrial sectors and their political allies as they are major stakeholders in setting and changing environmental standards. For example, in terms of the effluent standard, the VOCs have not been listed as examined items because of the disagreement from firms.[26] Being one of the major stakeholders, the industrial representatives from HSIP double as the setter of the "IC Industrial Air Pollution Control Standard," the "Overall Environmental Carrying Capacity Control on Air Pollution of HSIP," and the governing authority for issuing permits, while no environmental groups have been consulted.[27]

Even when there is a law to stipulate the development action or pollution problems, the environmental law is set aside. Development activities in HSIP are often protected by developmental policies such as SIP laws or the consensus of the Economic Development Advisory Conference (EDAC) and are exempted from environmental statutes. Although there are many environmental acts that could have regulated the firms in HSIP, few of them have been effectively implemented there. As discussed earlier in relation to the UMC EIA controversies, the violation of the EIA law did not result in the high-tech firm receiving any punishment. Instead, the controversies often end up quick passes of EIA review in order to mitigate business loss.

It is evident that the practices of the environmental authorities are marginalized by the SIP policy and further diminished by a crippled local autonomy. Although the local environmental agencies are authorized to investigate environmental practices in HSIP, they have been constrained by their governing capacity and uncoordinated institutional arrangement.[28] According to the former head of the Environmental Bureau of Hsinchu local authorities, there are only forty-eight and twenty-eight environmental personnel in Hsinchu City and County, respectively. The staff is in charge not only of the environmental investigation work in the HSIP, but also of all environmental work in the Hsinchu region (Kuo 2001; Liu *et al.* 2001). The disputes over the UMC 8-F EIA incident and the Quanta EIA incident further present the powerlessness of local authorities under the current

---

[26] The official raised his concerns: "The problem is, because there's no treatment plants to deal with waste solvent, the companies handle waste solvent by discharging it into the wastewater treatment plant that results in a high concentration of VOC in the wastewater of HSIP. It's almost impossible to examine one by one if we really want to," (interview with SIPA official).

[27] The consensus of the Economic Development Advisory Conference (EDAC), held in late August 2001, has contributed to the modification of many regulations to facilitate accelerating economic development. "Receiving many responses from the companies in the EDAC, the EPA suggested shifting the permit issuing authority from the Environmental Bureau to SIPA. This not only applied to the HSIP, but also applied to other industrial parks ..." (interview with SIPA official). The permit issuing authority for air emission and wastewater discharge was transferred to HSIP at the end of 2002.

[28] HSIP is located in the two jurisdictions of both Hsinchu City and County. It is not clear which agencies should be in charge of coordination when cross-boundary issues emerge.

institutional structure. A planning commissioner comments: "The companies can directly reach the Ministry of Economic Affairs (MOEA) and the President ... As a member of the local Urban Planning Committee, I find it difficult to raise opposition and everything seemed to be signed and passed. Although authority over land use has been transferred to the local government after the downsizing of the Taiwan Provincial Government, it's rarely implemented..."[29]

5.2. The Challenge of Environmental Carrying Capacity Control

Although the EIA law regulates development activities that aims at preventing and mitigating adverse impacts on the environment, the review of the EIA report is project-specific rather than a programmatic assessment that evaluates overall development activities in a region. There is no constant revision to assess the project's competing relationship with other development activities in the region, nor effective monitoring of implementation of EIA. According to Lu *et al.* (2002), the EIA system in Taiwan cannot accurately evaluate the impacts of an individual project on site due to insufficient data for comprehensive assessment of environmental carrying capacity in the region.[30] There is no programmatic and legislative EIA to assess long-term impacts of government policies.

Such an EIA system has failed to prevent or mitigate environmental impacts when facing rapid, agglomerated development driven by high-tech electronic industry. The environmental objective is set not on the basis of regional environmental carrying capacity, but on the basis of existing levels of pollution. For example, to manage the air quality, HSIP set the air pollution overall carrying capacity control standard to regulate the emission of air pollutants in the park. According to SIPA (2002), a "suitable" level was set, based on current pollutant emissions, for maximum emissions of air pollutants.[31] While the industry complained that the standard was not reasonable from the perspective of increasing production, the environmental community condemned the standard because it fails to address the problem of environmental carrying capacity, not to mention that community residents were not involved in the decision making processes.

The environmental communities have criticized the newly set standard as "drawing a bull's eye after shooting an arrow." It takes up all quotas of allowable emission without considering other point or nonpoint sources of pollution in the region. It is also worth noting that overall carrying capacity control is limited to air pollutants, and does not apply to water pollution, land use development, and water supply.[32] The policies, which promote industrial

---

[29] Interview with a Hsinchu City Planning Commissioner.

[30] The EIA reports are often criticized for a lack of factual basis and accuracy of data. The EIA's drawbacks in Taiwan also include lack of assessment of alternatives, failure to cover evaluation of clean production and product life cycle, lack of compensation and liability mechanisms for site cleaning, lack of cross-agency coordination, lack of bottom-up citizen participation, and lack of third party monitoring (Lu et al 2002).

[31] The standard accommodates current pollutant emission and leaves some room for future additional emission. For example, the estimated "suitable" emission level for VOC is based on the ratio of *increasing* VOC emission to the *increasing* rate of O_3 concentration. According to SIPA, when VOC emission doubles (compared to the current emission rate), the concentration of O_3 increases 1ppb, but when VOC emission increases threefold, the concentration of O_3 increases 6.4–14.4 ppb. Therefore, it suggested that maximum emission should not exceed twice the current emission (SIPA 2002).

[32] The government has hesitated to implement capacity control on water pollution because water flow (106,000 cmd) in the Kerya River is less than wastewater from HSIP (185,000 cmd) received into the Kerya River.

development, relax land use control, and prevent interference of environmental assessment and investigation to accommodate rapid IT development, conflict with long-term planning that seeks development and preservation of a sound environment.

5.3. A Lack of Active Citizen Participation in Environmental Processes

Participation of citizens and environmental groups has increasingly been included in the major pieces of environmental law passed since 1993. The Public Nuisance Dispute Resolution Act of 1992 partially incorporates citizens into the environmental decision making process. The Wildlife Conservation Act grants environmental groups a legal standing in participating in wildlife conservation affairs. The Environmental Impact Assessment Act includes more participatory procedures for citizens and environmental groups (Arrigo *et al.* 1996). The latest environmental legislation, the Environmental Protection Fundamental Act (2002) further grants citizens and interest groups the legal standing to sue the responsible agencies if the agencies fail to protect environment.

Although channels for public participation and formal access for the public to obtain governmental information are increasing, citizen participation in the environmental planning of pollution control and EIA review of development projects in HSIP has been limited. On the one hand, there is limited access for public participation in the planning and monitoring processes of HSIP. The formation of Environmental Monitoring Groups (EMG), in response to increasing grassroots demand for environmental investigation, provides a formal channel to involve environmental professionals, residents, and NGOs. However, participation in the EMG is confined to its invited committees rather than being open to the general public.[33] On the other hand, there is no significant organized force to demand more public participation from HSIP. Inactive citizen participation on social agendas is shaped by IT development characteristics, its economic driving force and the changes it effects in social demographic features. IT workers, who are also community members, are closely attached to the industry. Sharing a common interest with IT firms, they have kept silent on the environmental and health concerns associated with IT development, in the fear that a negative tone may undermine IT success and its glitzy image (Chang *et al.* 2004).

CONCLUSION

The failure of environmental practices is fully entrenched in political and economic power struggles. The state intervenes in political negotiations whenever there is a competitive use of resources that affects high-tech production. Enactment of many environmental laws and important amendments of older laws in the late 1990s does not guarantee sound

According to SIPA (2002), 185,000 tons of water discharge from HSIP per day and the related water quality examining items include BOD (maximum 25 mg/L), COD (maximum 80 mg/L), S.S (maximum 25 mg/L), which are all claimed to be lower than the national effluent standard.

[33] NGO representatives proposed to open up the EMG meetings to the general public and media, but the proposal was argued against by some of the committees. The official representatives are concerned the dialogues in the meetings may be misinterpreted by the media or that issues may get complicated when the public is involved. There was no conclusion as no consensus was reached (field note 2002/7/25).

environmental practice. The setting aside of environmental legislation, regulation, and implementation is evident in the face of high-tech development and the progrowth principle of "keeping industry rooted in Taiwan." The constraints of law enforcement are further embedded in the poorly designed institutional arrangement whereby local authorities and environmental agencies are not able to confront well-designed high-tech promotion policies. Two factors, weak environmental practices and strong state-led high-tech development strategies, act as firewalls to shield the high-tech sector from being environmentally responsible, and the governing entities from being politically responsible. Long-term national planning built upon the sustainable vision of resource use, land development, and industrial selection strategies yields to short-term rapid economic gain. The biggest restraints occur when community participation in environmental processes is not encouraged and mobilized for fear of retarding the industrial IT processes.

Nowadays, high-tech components are being built, assembled, used, and dumped around the world. The industry's unparalleled globalization has widely spread its economic influences, and many developing regions strive for and depend on the IT sector as it is viewed as an opportunity for leapfrogging. Taiwan's lesson demonstrates that as the IT industry grows, its toxic legacy and unquenchable demands for natural resources may pose unprecedented challenges to the existing environmental system. The power of IT dominance has also confined the opposition forces that aim at correcting its unsustainable developmental model.

The struggles in Taiwan present the pitfalls and limitations of greening and environmental leapfrogging. Yet Taiwan is not the only case. Many of the same problems can be found in other global high-tech communities in Scotland, India, Thailand, Japan, and China are revealed (SVTC and ICRT 2002; CAP 2002). There is a need to peel away high-tech fantasies, to examine and adjust governmental policies in favor of the high-tech sector through vast public investment and subsidy. Such retrospection and policy reform could not occur without vital grassroots movements that demand the corporation be accountable and the government be responsible, for the sake of social justice and a safe, clean environment. As the governance of high-tech economy is still under construction, active campaigns in pursuing sustainable high-tech development have the potential to influence the shape of construction. The emergence of a transnational environmental justice network that aims to correct unsustainable models of IT development and campaigns for clean production, corporate social accountability, and effective civic engagement may offer a new venue to address the issues at root.

References

Arrigo, LindaGail, Tze-Luen Lin, and Yvonne M. Lin. 1996. "Environmental Conditions and Environmental Law in Taiwan." *A World Survey of Environmental Law*. Capria, Antonella *et al.*, eds., pp. 765–777. Milano, Italy: Gluffre Editore.

BAN (Basel Action Network). 2002. Exporting Harm. The High-Tech Trashing of Asia. http://www.svtc.org/cleancc/pubs/technotrash.pdf.

CAP (Corporate Accountability Project). 2002. *Beyond Good Deeds: Case Studies and a New Policy Agenda for Corporate Accountability*. Chapter 3. California: Nautilus Institute, http://www.nautilus.org/cap/BeyondGoodDeedsCSRReportNautilusInstitute.pdf.

Castells, Manuel and Peter Hall. 1994, *Technopoles of the World: The Making of 21st Century Industrial Complexes*. NY: Routledge.

Chang, Shenglin. 2000. Real Life at Virtual Home: The Silicon Landscape Construction in Response to the Transcultural Home Identities. Dissertation. Department of Landscape Architecture and Environmental Planning, University of California at Berkeley, Berkeley.

Chang, Shenglin, Hua-mei Chiu, and Wenling Tu. 2004. "The Silence of Silicon Lambs: Speaking out Health and Environmental Impacts within Taiwan's Hsinchu Science-Based Industrial Park" In *Proceedings of 2004 IEEE ISEE Symposium/Summit*. pp. 258-263.

Chang, Shenglin, Wen-Ling Tu, Wenchun Yang, Li-fang Yang. 2001. *A Study of the Environmental and Social Aspects of Taiwanese and U.S. Companies in the Hsinchu Science-based Industrial Park*. Working paper for Nautilus Institute for Security and Sustainable Development.

Chen, Hui-Min. 2002. "De-construction of the Environmental Myth on HSIP and High-tech Industries." Masters thesis. Journalism Graduate School, National Chengchi University, Taipei, Taiwan.

Chen, Leon-Jiun. 2001. "High-Tech Fantasies: Hsinchu Science-Based Industrial Park and Local Development, Master Thesis. Department of Geography, National Taiwan Normal University, Taipei, Taiwan.

Chen, Kuan-Fu. 1990. "Dependent Development and Spatial Structure of Taiwan High-Tech Industries: A Case Study of HSIP." *Taiwan: A Radical Quarterly in Social Studies*, 3 (1): 113–149.

Chien, Shu-Shen. 1997. "State, Development Policy, and Planning on Special Zone: A Case Study of Development and Transformation of Taiwan Exporting Zone." Masters thesis. Building and Planning Graduate School, National Taiwan University, Taipei, Taiwan.

Chiu, Hwa-Mei. 2000. "Environmental Crisis Seizes the IT Industry's Throat." *Common Wealth Magazine*, (228): 112–120.

Environmental Bureau of Hsinchu City Government. 2000. *Planning Report: Hsinchu Science-Based Industrial Park Air Quality Protection and Pollution Control Strategies*. Hsinchu, Taiwan.

Evans, Peter. 1997. The Eclipse of the State? *World Politics*, 50 (1): 62–87.

Foran, Tira. 2001. "Corporate Social Responsibility at Nice Multinational Electronics Firms in Thailand: A Preliminary Analysis." Working paper for Nautilus Institute for Security and Sustainable Development, Report to the California Global Corporate Accountability Project.

Gopalan, Radha. 2000. "A Study on the Performance of the Indian IT Sector." Working paper for Nautilus Institute for Security and Sustainable Development, Report to the California Global Corporate Accountability Project.

Hall, Peter. 1998. *Cities in Civilization: Culture, Innovation, and Urban Order*. Chapter 14. Phoenix, AZ: Orion Books.

Hsia, Chuang-Wei. 2000. "How Many Science Parks are needed in Taiwan." *Common Wealth Magazine* (228): 103–111.

Jussawalla, Meheroo. 2003. "Bridging the Global Divide." *Information Technology Parks of the Asia Pacific*, M. Jussawalla and R. Taylor. Armonk, NY: M.E. Sharpe, eds. pp: 3-24.

Kao, Da-Jen.1995. "A Comparative Study of Development Experiences of Taiwan Exporting Zone and HSIP." Masters thesis. Building and Planning Graduate School, National Taiwan University, Taipei, Taiwan.

Ku, Yang. 2002. "Environmental Consideration and Management Strategies for High-technology Parks." In *Proceedings of the Conference of the Hsinchu City Sustainable Development Model Plan: Green Silicon Island International Environmental Protection Technology Exchange.* Hsinchu, Taiwan, pp. 46-58.

Kuo, Kun-Ming. 2001. "Environmental Protection in Hsinchu County: Current Situation and Constraints." In *Proceedings of Sustainable Development – Green Silicon Island Environmental Potentiality Conference.* Hsinchu, Taiwan, pp. 11-16.

Lin, Si-Kai. 2002. "News Analysis on Taiwan's Industries and Environmental Protection: A Case Study of News Report on HSIP." Masters thesis. National Chengchi University, Taipei, Taiwan.

Liu, Chia-Chun, Kuan-Wei Liu, and Yu-Feng Liu. 2001. "HSIP Current Situation and Future direction for Environmental Control: Report from Environmental Bureau of Hsinchu City." In *Proceedings of Sustainable Development- Green Silicon Island Environmental Potentiality Conference.* Hsinchu, Taiwan pp. 1-10.

Lu, Kuo-Shen, Weita Fang, and Li Sheng. 2002. "Discussion on Direction of EIA System Development in Taiwan." National Policy Foundation Report. Taipei, Taiwan.

Luthje, Boy. 2002. "The Detroit of the New Economy: The Changing Workplace, Manufacturing Workers and the Labor Movement in Silicon Valley." Paper presented at the Annual Convention of the American Historical Association. 1/6/2002, San Francisco, CA.

Mathews, John A. 1997. "A Silicon Valley of the East: Creating Taiwan's Semiconductor Industry." *California Management Review*, 39 (4) Summer. Pp. 26-54.

Mazurek, Jan. 1999. *Making Microchips: Policy, Globalization, and Restructuring in the Semiconductor Industry.* Cambridge, MA: MIT Press.

Saxenian, Anna Lee. 2001. "Taiwan's Hsinchu Region: Imitator and Partner for Silicon Valley." SIEPR Discussion Paper Series reports, Stanford University.

— 1999. "The Silicon Valley–Hsinchu Connection: Technical Communities and Industrial Upgrading." SIEPR Discussion Paper no. 99-10, Sept. 1999, Stanford Institute for Economic Policy Research.

— 1996. *Regional Advantage: Culture and Competition in Silicon Valley and Route 128.* Cambridge, MA: Harvard University Press.

— 1991. "Institutions and the Growth of Silicon Valley." *Berkeley Planning Journal*, 6, pp. 36-57.

SIPA. 2002. *Hsinchu Science-based Industrial Park Environmental Impact Assessment Report: Investigation Analysis and Strategies.* Hsinchu, Taiwan.

— (1997). "Report on the Issues of Media Inquiry on HSIP Wastewater Pollution." 11/3/1997.

SVTC and ICRT. 2002. "Conference Report: Workshop on Global Strategies for a Sustainable High-Tech Industry." 11/14–17/2002. San Jose, CA.

Wangel, Arne. 2001. "Manufacturing Growth and Social Deficits: Environmental and Labor Issues in the High Tech Industry of Penang, Malaysia." Working paper for Nautilus Institute for Security and Sustainable Development, Report to the California Global Corporate Accountability Project.

Wheeler, James O., Yuko Aoyama, and Barney Warf (eds). 2000. *Cities in the Telecommunications Ages: The Fracturing of Geographies*. New York, NY: Routledge.

NEWSPAPERS

China Times, 2002/2/28. Lee, Chen-Wei. "Drought! HSIP is really scared this time," 5.

China Times, 1998/11/11. Chen, Chuon-Sin. "Park's wastewater release pungent odor, KaoFong Li's residents hold breadth, some residents have nausea and seek treatment," 8.

Industrial and Business, 2002/2/28. Lu, Se-Hui and Shu-Chen Tan. "Lin, Hsin-Yi announce fallow plan to solve the problem of water shortage in HSIP," 3.

Taipei Times, 2004/9/3. Staff writer with DPA. "Water cuts cause massive losses," 10.

Taipei Times. 2002/12/12. Chiu, Yu-Tzu and Ko Shu-Ling. "Shortage of rainfall leaves fields fallow," 1.

Taipei Times. 2002/5/3. Staff writer. "Shortage of water to hit production," 17.

Taipei Times, 2001/8/17. Lin, Chieh-yu and Kevin Chen. "Chen `on his knees' for less red tape," 1.

United Daily. 2005/10/19. Ting, Wan-Ming. "HSIP's firms refuse construction plan of desalinator," 4.

United Daily. 2000/7/21. Lee, Ching-Lin. "Treatment price for waste solvent is going high, SIPA plans to build incinerator to turn the waste solvent into fuel," 4.

In: Leapfrogging Development in Emerging Asia
Editor: Peter Ho, pp. 129-149

ISBN: 978-1-60456-410-5

Chapter 7

ENVIRONMENTALISM IN TAIWAN. A CASE OF EMBEDDED AUTONOMY AND GREENING?[1]

Li-Fang Yang
Sociology Department at University of Wisconsin-Madison

1. INTRODUCTION

In the late 1980s, there was a significant shift in the character of environmental crises, from intrinsically local issues to global issues, such as global warming and ozone depletion. Corresponding to such a shift, there has been a renewed interest in environmental issues among the general public, policy makers, environmentalists, academics, and business decision makers. This new wave of the environmental movement embraced policy innovations (e.g., market-incentive environmental regulations), technological innovation (e.g., industrial ecology, life-cycle analysis, environmental management and ISO 14000), and pervasive public discourse (e.g., sustainable development). Against this backdrop, writings on Greening and Ecological Modernization Theory (EMT) emerged in Europe. The latter can be regarded as a social change theory, devoted to conceptualization of the institutional changes pertaining to environmental problems and solutions.

Ecological Modernization Theory distinguishes itself from a critical approach, and suggests fundamental and constructive roles that science, technology, capital and the state can play in the process of environmental improvements. Its central thrust is that environmental problems have been caused by modernization and industrialization, and that their solutions must necessarily lie in modernization and "superindustrialization" rather than de-industrialization. Emerging from the context of European industrial advanced societies, EMT witnesses the institutional transformation in process – radical restructuring of production,

[1] This chapter was prepared for the conference 'Managing the Environment in Developmental States: Asian Perspective,' sponsored jointly by Tsinghua University, Beijing, PR of China and the Environmental Policy Group, Wageningen University, The Netherlands, held at Tsinghua University, Beijing, PR of China on Sept. 26–27, 2002.

consumption, state practices, and political discourses along ecological rationality (Hajer 1995). However, as Buttel (1999) has argued, due to the fact that the development imperatives, the political culture, the state structure, and the social institutions differ across countries, the environmentalism, its mechanisms, and the institutions that back up environmental policies or environmental reforms differ accordingly. Consequently, the legitimacy of EMT in explaining environmental reforms in less industrialized countries and newly industrializing countries becomes a contested question.

This chapter contributes to this discussion by examining the usefulness of EMT concepts for evaluating and shaping environmental improvements in newly industrializing countries, taking as one case Taiwan, specifically Taiwan's high-tech industries. High-tech industries are the most globalized, innovative, and strategically important industries in Taiwan's economic development plans. Hence, we are most likely to detect ecological modernization in Taiwan's high-tech industries if ecological modernization is happening in Taiwan at all. Further, the latest competitive shift in the Taiwanese industrial structure has been away from high-pollution intensities toward low-pollution intensities, and high-tech industries are promoted under the statute for environmental consideration. The nexus between the state and high-tech industries manifested itself as the best social institution for understanding the specific dynamics of environmental reforms.

Taiwan is known, in the political economy literature, as a developmental state, in which the state plays a leading role in planning and prioritizing economic development in the process of industrialization. The rapid economic development track that Taiwan has been able to stay on drew many scholars' interests in studying the effectiveness of Taiwan's economic structure and the synergy between the state and the market, i.e., the embedded autonomy (Evans 1995, 1997; Ho and Edmonds 2007). Given its unique economic development experiences and state-economic structure, Taiwan's environmental reforms, particularly economic and environmental development in high-tech industries, could provide a valuable empirical contribution to EMT debates. This chapter examines Taiwan's ecological modernization process and dedicates the results to the furthering of EMT.

1.1. Purpose of the Research

Two main questions are addressed in this study. First, the extent to which environmental reform is taking place in Taiwan in general and in high-tech industries in particular, and what mechanisms bring about the reforms. Second, using EMT as a theoretical perspective, the author intends to contribute to theoretical discussion of EMT by examining the usefulness of EMT concepts for evaluation and shaping environmental protection efforts in newly industrializing countries – specifically Taiwan's high-tech sector.

In the next section, the framework of EMT will be summarized to aid our analysis. In section three, an overview of Taiwan's environmentally related institutional changes will be provided, the role social agents play in this transformation will be discussed, and the extent to which these changes cause ecological concern will be addressed. In the fourth section, the role the state plays in the development of high-tech industry is detailed, followed by how this relationship of embedded autonomy between state and high-tech industries facilitates the implementation of environmental policy. This chapter will conclude with discussions, in terms of the social, political, and economic spheres of EMT, on the trends and challenges of economic transition and ecological modernization in Taiwan.

2. Research Framework

For the purpose of examining the applicability of EMT in Taiwan's environmental reform efforts, it is necessary to identify the key characteristics of EMT. In the early development of EMT, scholars emphasized the technical aspect of environmental improvement in the process of production and in the conservation of material (Janicke *et al.* 1989; Hajer 1995). Clean technology, recycling, waste reduction, dematerialization and renewable resources are identified as the path toward clean production. Another focus for EMT scholars has been identifying the institutional and social aspect of environmental transformation. Scholars have suggested (Spaargaren 1997; Sonnenfeld 2000) a few significant institutional characteristics for ecological restructuring in Western Europe. The state regulatory structure has shifted from centralized top-down management to decentralized participatory negotiation management regarding environmental policy.

The roles that nongovernmental organizations play has moved from outside to inside environmental policy networks, and direct interaction between business and NGOs has become a new trend. A summary of institutional characteristics is provided in Table 1.

Table 1. Institutional transformation for ecological modernization

1. Science and technology are not only causes of environmental problems but also valuable and potential sources of solutions.
2. Market dynamics and economic agents are not just the disruptive force of ecological degradation, but are the carriers of ecological restructuring and reform.
3. The role the nation-state plays in environmental reform has changed from top-down, centralized command and control environmental regulation to more decentralized, flexible and consensual styles of governance.
4. Social movements are more engaged in public and private decision-making institutions regarding environmental reform rather than merely protesting.
5. There are changing discursive practices and emerging new ideologies regarding environmentalism with a vision of sustainability.
6. Science and technology are not only causes of environmental problems but also valuable and potential sources of solutions.
7. Market dynamics and economic agents are not just the disruptive force of ecological degradation, but are the carriers of ecological restructuring and reform.
8. The role the nation-state plays in environmental reform has changed from top-down, centralized command and control environmental regulation to more decentralized, flexible and consensual styles of governance.
9. Social movements are more engaged in public and private decision-making institutions regarding environmental reform rather than merely protesting.
10. There are changing discursive practices and emerging new ideologies regarding environmentalism with a vision of sustainability.

In addition to the technical and institutional aspects of ecological modernization, a third focus of EMT has been on mechanisms of change. Two mechanisms are seen as operative. The first mechanism, "economizing the ecology," fosters internalizing previously externalized costs by ascribing monetary value to natural resources. The second mechanism, "ecologizing the economy," embeds the economy into ecology, and incorporates knowledge of natural cycles into economic activity.

3. An Overview of Taiwan's Environmental Reforms

The Emergence of Environmentalism

The emergence of environmental awareness has its own roots in the reaction to the outbreak of environmental harms (Hsiao *et al.* 1995). According to four waves of national survey work on public awareness of Taiwan's environmental problems conducted by Hsiao (1982, 1986, 1989, 1991), he argues that Taiwan's environmental movements and politics can be characterized by three distinctive streams: the antipollution protests, dating from 1980; the nature conservation movement, begun in 1982; and the antinuclear movement, which started in 1988 (Hsiao 1999: 32). Antipollution has awakened Taiwan's environmentalism since 1980, and it has continually been the major collective action pertaining to environmental protection. Antipollution protests were mostly mobilized against, or sought compensation for past injuries or losses to residents. The movement was reactive rather than preventive or preemptive. Preventive or preemptive protests accounted for only 16 percent of the 1,211 local antipollution protests that took place between 1980 and 1996. As Hsiao (1997) argues, this stream, the dominant one, of Taiwan's environmentalism can be characterized as victim activism.

In the early stage, Taiwan's political institutions disadvantaged the development of an environmental movement. Under the one national party rule, all kinds of social movement confronted substantial difficulties in mobilizing social resources to draw on appropriate support and to legitimize their actions. However, as a result of political liberalization, educational pervasiveness and advancement, and the development of mass media since the 1980s, the environmental movement has been able to attract more attention from the general public and stimulate greater environmental awareness. In particular, after the lifting of martial law and other long-time political bans in 1987, the frequency of environmental protests increased significantly. According to Hsiao's (1997) study indicated above, of the antipollution protests that occurred between 1980 and 1996, over 90 percent of 1,211recorded disputes took place between 1988 and 1996.

Besides the occurrence of quantitative changes in environmentalism in Taiwan, there have gradually been qualitative changes. In the past decade, the scope of environmental awareness in Taiwan has no longer been limited to the issues of environmental protection and ecology preservation at the local level. Instead, it has been enlarged and participates in the global discourse regarding issues of sustainable development, including global warming, biodiversity, and environmental justice. Those involved in environmental movements are coming from more social classes, from local victims to scholars, writers, students, scientists, aboriginals, and politicians.

3.1. Democratization and the Proliferation of NGOs

Some environmental literature presumes that democratization is a prerequisite for the progress of environmentalism. However, as Ho and Edmonds have demonstrated it can be a contested question (Ho and Edmonds 2007), and Taiwan is a case in point. It is noted that the early development of environmental movements and other social movements has brought in democratization processes as a result of their efforts. The rapid increase in environmental protests in Taiwan reflects not only a direct response to the deteriorating living environment and ecosystem, but also an activist culture that encourages people to take direct action to solve society's environmental problems during the political liberalization period of Taiwan (Yang and Lin 2000). In the early 1980s, labor, farmer, women, and student movements, either based on sympathy or strategic concerns, came into coalition with one another to mobilize necessary social resources in order to challenge the long-standing authoritarian state. We could easily find overlap of leadership in these social movements. These coalitions promulgated the proliferation of nongovernmental organizations in Taiwan, especially after 1987.

These organized civil groups play significant roles in the development and evolution of Taiwan's environmental movement. By one estimate there were over 232 environmental NGOs in Taiwan (Hsiao 1997b). Major groups, including the Taiwan Environmental Protection Union (TEPU), the Taiwan Greenpeace Association[2], the Ecological Conservation Alliance, the Wildbird Society, the Homemaker's Union and Environmental Protection Foundation, the Life Conservationist Association, and the Chung-Hua County Pollution Control Association, were actively attempting to participate in and influence environmental policy making locally and nationally, largely through demonstrations, elections, coalitions, and information (Lin 1995). Interestingly, there have been no international environmental NGOs established in Taiwan. While violence has occasionally occurred at environmental protests, there was no evidence showing that environmental NGOs in Taiwan have intentionally adopted "violence" as a tactic to influence decision-making. Although these organizations do not define the entirety of the environmental movement, they are clearly the most visible and often the most influential actors in environmental policy debates, as well as the catalysts for the construction of environmental awareness in Taiwan (Yang and Lin 2000).

Although Taiwan's environmental NGOs have grown in number, difficulties in recruiting members and looking for financial support have also emerged. Another weakness that Taiwan's environmental NGOs have shown is a lack of information resources from international environmental NGOs. To connect with international environmental NGOs, providing sophisticated environmental knowledge to handle high-tech environmental problems, is greatly needed by Taiwanese environmental groups.

3.2. The Industrial Structure and the Environment

Rapid industrialization and modernization have long been blamed for the ecological devastation in contemporary societies. With a double-digit growth rate in economic development over the past thirty years, Taiwan's "miracle economy" has brought material

[2] The Taiwan Greenpeace Association is an independent entity, which has no association with Greenpeace.

wealth to its people, but in the process created ecological hardships for the island to bear. As the public became more aware of the high environmental risks incurred by rapid economic development, business became more aware of the environmental costs incurred as a result of rigid environmental regulations, and the government was frequently confronted with environmental management challenges. At the same time, as the world's thirteenth largest trading economy, Taiwan also experienced structural change in its economic sectors. The production value of the industrial sector grew from 19.7 percent in 1952 to 47.1 percent of Taiwan's GDP in 1986, and then gradually dropped to 27.7 percent in 1997; while the agricultural sector shrank from 32.2 percent to 2.7 percent during the same period. Meanwhile, the production index of the heavy chemical industry increased 2.3-fold from 1987 to 1997 (CEPD 1999). This changing industrial structure carries considerable implications for environment policy and resource uses in Taiwan.

3.3. Structural Embeddedness of Environmental Problems: The Developmental State, Embedded Autonomy, and Flexible Capitalism

In the political economy literature, Taiwan's development experience has been widely regarded as the typical model of a "developmental state," (Amsden 1985; Deyo 1987; Wade 1990). This refers to states that embrace economic development as the foremost goal of state action and play a leading role in the process of capital accumulation. The role of the state in promoting economic growth is to guide the direction of the economy through selective state intervention, state industrial planning, and development strategies and policies. As Castells points out, "The core of Taiwanese economic growth lies in the notion of flexible production ... The flexibility concerns both the industrial structure itself, and the adaptability of the overall structure to the changing conditions of the world economy, under the guidance of an all-powerful state ..." (Castells 1992: 41).

In fact, the core of such "flexible capitalism" lies in the Kaomintang (KMT) state and the highly diversified networking relationships among business groups, between the state and business, and between businesses and the world economic system (Castells 1992: 42). This "flexible capitalism" has constructed an important dimension of Taiwan's ecological crisis – uneven development and environmental pollution. The companies with close ties to the state are prone to use such relationships to resolve environmental disputes instead of taking on some corrective action. The misuse of such state–business relationships is prevalent.

3.4. Institutional Change in State Economic Development Plans: Away from Traditional Energy-Intensity and Toward High-Tech Industries

Chemical, petrochemical, paper and pulp factories, and steel mills are perceived as the most polluting and energy-intensive industries. These energy-intensive industries accounted for approximately one-third of Taiwan's total energy consumption, while contributing only 6.69 percent of GDP in 1997 (Wang, 1998). It has been argued that inadequate industrial structure and energy policies resulted in serious environmental pollution and increased CO_2 emissions in Taiwan. The idea to shift the industrial structure toward a high-tech, less-polluting, information-based economy was becoming the consensus among policy makers and environmentalists (MOEA 1999; Wang 1998).

Since the 1980s, the state of Taiwan has begun to focus on the strategic development of high-tech, high value-added, and energy-efficient industries. The establishment of the

Hsinchu Science-Based Industrial Park (HSIP) in 1980, administered by the National Science Council, was a case in point. Through tax credit and investment benefits, businesses were encouraged to intensify research and development to enhance competitiveness in the world economy. By 1990, high-tech products, mainly electronics, information and machinery products, accounted for 40.2 percent of total exports. In 1991, based on the Six-Year National Development Plan, ten emerging high-tech industries, including semiconductors, communications, pollution control and treatment industries, etc., were selected as development goals for the next phase of Taiwan's development. It was expected that the proportion of the technology-intensive industries in overall production would increase to 40 percent by the year 2002 (IDB 1998). Six selection principles for the ten emerging industries were: 1) great market potential; 2) high linkage with other industries; 3) high added value; 4) advanced technical level; 5) low pollution; 6) less dependence on energy supplies (IDB 1998).

Environmental consideration had begun to be integrated into economic plans and industrial policy.

3.5. High-Tech Industries and the Environment

Environmental problems within the semiconductor industry in Taiwan were not well documented until 1997. Previously the high-tech industry constructed a clean and nonpolluting image, free from environmental and occupational hazards. A fire accident at United Integrated Circuits sounded the alarm that there were potential environmental risks that accompany high-tech industries. People witnessed a huge amount of chemicals and toxic materials being emitted into the air, and serious safety and environmental concerns about the semiconductor industry caught people's attention for the first time in Taiwan (Yang and Lin 2000). The environmental risk in the semiconductor industry is not readily resolved by end-of-pipe pollution control measures.

It is not widely known that the semiconductor industry uses a variety of highly toxic materials in the process of production during cutting, polishing, dicing, and packaging. Also, the semiconductor industry changes its processes, chemicals, and technologies rapidly to enhance its global competitiveness so that the process of manufacturing is difficult to fully understand. In addition to the material hazards, the character of high energy and high water consumption in the semiconductor manufacturing process has also raised environmental impact concerns.

Table 2. Demand for water, electricity and land for IC production by 2010

	1995	2000	2005	2010
IC production value	1,000	11,600	23,200	35,000
Water (CMD)	22,000	255,200	510,400	770,000
Electricity (KW)	112,000			1,400,000
Land (hectare)	36.8	427	854	1,288

Source: IDB 1996.

3.6. Environmental Policy and the Regulatory Framework

Given the growing environmental pressure, Taiwan's government was forced to react to environmental movements initiated by local civil society and the international community. In reaction to local antipollution protests, tougher pollution controls were adopted. For instance, a specific Pollution Disputes Resolution Law was passed, the Environmental Impacts Assessment Act was enacted, and other environmental protection laws were passed. Historically speaking, the environmental regulatory framework began forming in the 1970s and has developed rapidly since 1987, when the Taiwan Environmental Protection Administration (Taiwan EPA) was established under the Executive Yuan. The Taiwan government passed its first comprehensive environmental policy plan, *Guidelines for Environmental Policy at the Current Stage*, in October 1987, shortly after the Lukang residents' campaign against government approval for a titanium dioxide plant to be built by Dupont in their community (Yang and Lin 2000).

The administrative authority was comparatively weak in the Taiwan government. The weakness of administrative authority in the Taiwan EPA mainly results from the fact that the tasks of managing Taiwan's environmental affairs are distributed among various jurisdictions of governmental departments and agencies. Environmental affairs are divided into two major regulatory and administration systems: pollution control and regulation on the one hand, and natural resource management on the other (Chiou 1995). Responsibilities for natural resources management and environmental protection are distributed among several central governmental departments and agencies, such as the Ministry of the Interior (national parks services and land uses), the Council of Agriculture (forest, natural conservation and wildlife protection), and the Atomic Energy Council (nuclear power regulation and radiation protection). In contrast, the major responsibilities for the Taiwan EPA are environmental pollution control, setting standards, and enforcement. The barrier to communicating and coordinating environmental affairs among governmental agents casts a long shadow on the prospect of Taiwan environmental management.

Currently, more than 106 laws and regulations have been promulgated, covering air, water, noise, soil and groundwater pollution, recycling, solid and hazardous waste, and public nuisance disputes. Legislation is pending in the areas of marine pollution control and the Environmental Protection Basic Act, which was regarded as the "environmental constitution" of Taiwan. Environmental and ecological protection is written into the Constitution as deserving equal priority and effort with economic and technological development. The high prioritization of environmental projects in the Six-Year Development Plan, the Industrial Development Bureau's (IDB) five-year environmental improvement plan, and the Taiwan EPA's National Environmental Protection Plan have all attested to putting teeth in environmental regulation and building capacity on the island to deal with environmental problems.

The Environmental Impact Assessment Act of 1994 is the most important environmental legislation and includes more participatory procedures for citizens and environmental interest groups. Environmental impact assessment is required for all major development projects. The Act requires that the authorities establish a review committee and that it should be composed of not less than two-thirds scholars and environmental professionals, and gives the Taiwan EPA the right to veto projects that have an adverse impact on the environment.

Overall, Taiwan's environmental regulatory framework has shifted from low-level regulation in the 1970s (Huang 1994), via the command-and-control strategy of the 1980s, to an array of market-oriented systems. The Taiwan EPA has initiated permit systems for air and water effluents. The system will eventually track 17,999 stationary sources of air pollution and 12,000 point sources of industrial wastewater. Air pollution control fees were also introduced in 1997.

Table 3. Air pollution control standards for semiconductor industry

Air pollutants	Emission standards
VOCs	Emission reductions greater than 90% or total factory emissions less than 0.6 kg/hr (according to methane calculation basis).
Tricholoroethylene	Emission reductions greater than 90% or total factory emissions less than 0.02 kg/hr.
Nitric acid, hydrochloric acid, phosphoric acid, HF acid	Emission reductions greater than 95% or total factory emissions less than 0.6 kg/hr.
Sulfuric acid	Emission reductions greater than 95% or total factory emissions less than 0.1 kg/hr.

Source: EPA 1999.

The air pollution emission standards for the semiconductor industry were announced in 1999. These standards, which are based upon emission quantities and reduction amounts, have been effective since July 1, 2000. The new standards define and cover the semiconductor industry comprising integrated circuit wafer manufacturers, wafer packaging firms, wafer-stacking companies, semiconductor masking firms, and circuit frame manufacturers. Targeted pollutants include volatile organic compounds (VOCs), trichloroethylene, nitric acid, hydrochloric acid, phosphoric acid, hydrofluoric acid, and sulfuric acid.

According to the Taiwan EPA, semiconductor firms consumed 11,500 tons[3] of VOCs annually, including benzene, methylbenzene, isopropyl alcohol, dichloroethylene, and trichloroethylene, resulting in approximate 3,000 tons of emission per year. The new regulation requires VOC emissions to be reduced by 90 percent. In addition, about 600 tons of inorganic acids were emitted each year by the semiconductor industry, which current environmental regulations have not yet effectively controlled.

Under the above-mentioned increasingly sophisticated environmental policy, Taiwan's environment has seen some visible improvements. For instance, Taiwan's government intended to comply with the Montreal Protocol in 1989 and began limiting the import of CFCs, which were widely used by the electronics industry and joined the efforts of local industrial, academic, and research sectors in searching for CFC alternatives. These efforts resulted in reducing CFC consumption in Taiwan from 10,159 tons in 1986 to 2,493 tons in 1995, a 75 percent reduction over 1986's baseline consumption. This figure has continued to drop since then. Taiwan's electronic industry has adopted CFC alternative technologies, such as HCFCs, no-clean, aqueous, and semi-aqueous processes.

[3] Units for Taiwan EPA data are metric tons.

Taiwan also achieved notable results in its air pollutant emission reduction efforts. According to the Taiwan EPA, statistically, even though the "natural growth emission"[4] is around 5 to 10 percent in Taiwan, total pollution emission has decreased for most air pollutants according to annual trends of air pollutant emissions. During the period 1991 to 1998, measured levels of PM10 were reduced from 389,000 tons to 288,000 tons, a 26 percent reduction. In the same period, SOx was reduced from 604,000 tons to 374,000 tons, a 38 percent reduction; NOx was reduced from 517,000 tons to 486,000 tons, a 6 percent reduction; HC was reduced from 946,000 tons to 810,000 tons, a 16 percent reduction; CO was reduced from 3,102,000 tons to 1,826,000 tons, a 41 percent reduction; Pb was reduced from 591 tons to 261 tons, a 56 percent reduction. The emission reduction shows state-promoted air pollution control strategies, including emission standards, pollution behavior control, and economic-incentives, have been successfully implemented.

In the preliminary stage of the EPA's establishment, 1987–1991, the percentage of unhealthy air quality station days was about 16 percent. After the Air Pollution Control Act was enacted in 1992, the EPA has actively promoted pollution control methods for industries and vehicles.[5] In 1997, the percentage of unhealthy air quality station days was reduced to 5.46 percent.[6] In 1998, the percentage of unhealthy air quality station-days was further reduced to 5.09 percent. And according to the stipulated targets of the National Environment Production Plan, the percentage of unhealthy air quality station days will be reduced to 3 percent in 2001, 2 percent in 2006, and 1.5 percent in 2011 (EPA 2002, http://www.epa.gov.tw).

In addition to air quality improvement, water quality improvement has also been effective. In 1999, 0.37 percent of the total tap water samples taken were found to be below minimum quality requirements, compared to 0.41 percent in 1998, 0.9 percent in 1997, and 2.95 percent in 1996.[7]

3.7. State-Assisted Environmental Management

Although environmental groups were critical of the environmental regulations and standards set out by the Taiwan EPA, many companies began to invest in environmental pollution equipment and improve processes for cleaner production. The Taiwan EPA, the IDB and the Industrial Technology Research Institute (ITRI) have been leading and developing programs to assist companies in establishing environmental management systems. Through the amendments to the Statute for Encouragement of Investment, tax credit and investment incentives were provided to industries for improving pollution control, energy efficiency and conservation, recycling, and waste reduction. For example, manufacturers are eligible for 5 to 20 percent of a company tax credit for expenditure on environmental protection equipment or energy conservation technology. Anti-pollution investment plans, or construction projects are qualified for low-interest loans. An Industrial Pollution Prevention Technology Advisory

---

[4] Population growth, industrial development, and economic activity themselves increase air pollutant emissions. The increment of increasing air pollution emission due to such growth, so-called "air pollutant natural growth emission," is unavoidable in social development processes. Pollution controls are mitigating this growth-related momentum in emissions.

[5] The main air pollutant emission sources in Taiwan are stationary sources (factories and construction sites) and mobile sources.

[6] By comparing the percentage with those in 1986 and 1987, the rates of progress were 15 percent and 68 percent, respectively.

[7] The 1998 figure was the lowest in fifteen years.

Task Force was also established by IDB to assist industries to reduce pollution and minimize waste in the most cost-effective manner.

For both economic and industry safety concerns, many of the semiconductor companies in HSIP are making efforts to address environmental issues. It was estimated that over 50 percent of the companies in HSIP have been ISO 14000 certified, including the leading semiconductor companies in Taiwan, United Microelectronics Corporation (UMC), Taiwan Semiconductor Manufacturing Company, and Winbond Electronics. For UMC alone, its investment in pollution control equipment exceeded NTD50 million in 1999. It is estimated that environmental practices and resource conservation resulted in economic benefits of NTD537 million (UMC 2000).

In short, state-assisted environmental management has characterized the industrial and environmental policies in Taiwan. With policy goals of maintaining sustainable economic growth, improving living standards, and protecting the environment, the state has not only assisted firms to make structural adjustments and accelerate upgrading to cleaner production, but has also adjusted industrial structure and the trajectory of technical modernization.

The following section explains how the embedded autonomy relationship between the state and high-tech industries facilitates the implementation of environmental policy.

4. Taiwan High-Tech Industries

4.1. The Role of the State in Developing High-Tech Industries

The state's involvement in the development of the semiconductor industry is considered vital for success. In the 1970s, when the semiconductor industry was in its infancy, requiring intensive capital investment for its research and development, the state's incubating role become critical in providing technical,[8] financial, and institutional support for its success. For building semiconductor industries in Taiwan, the Industrial Technology Research Institute (ITRI) and the Electronic Research Service Organization (ERSO) were established as main institutions to cultivate the technologies, rapidly diffuse technologies, and give birth to the semiconductor industry. Major private firms such as United Microelectronics, The R.O.C.'s Economic Development Plan Electronics Corporation and Taiwan Semiconductor Manufacturing Company were created directly under the state's sponsorship. The state provides virtually everything needed for the formation of major semiconductor manufacturing firms, including capital investment, technology transfer and manpower (Mathews and Cho 2000).

[8] In the 1970s, Taiwan's entry into the semiconductor industry was considered late, especially for a state lacking in key areas of technological capacity. Beginning in the 1970s, Taiwan made successful inroads into the semiconductor industry by directly importing key technologies developed by the United States and Japan. The Taiwan government launched its first semiconductor project in 1974. In September 1982, the information industry was selected as a strategic sector in the state's developmental plans, and has since received a string of investments from the government.

4.2. Social Network between Public and Private Sector

The social network in Taiwan's high-tech industry presented itself as a unique structure. In fact, many major semiconductor firms were spinsoffs of ITRI/ESRO, which not only transfer technology but also provide human resources for the new companies. Most of the important figures in Taiwan's semiconductor industry, both in the government and the private sector, began their careers in these two organizations. Personal ties, based on common professional background, also provided a trust-based relationship for linking firms in the production networks.

4.3. The High-Tech Cultivator: The Industrial Technology Research Institute

The Industrial Technology Research Institute (ITRI) is one of the most important innovative institutional devices that the Taiwanese state has employed to direct the path of industrialization. In the case of the semiconductor industry, the scale and scope of the industry in Taiwan expanded on its secure foundation of the technological capabilities acquired within ITRI/ERSO, and diffused to a small number of firms. Mathews and Cho (2000) comment on the remarkable success of this institutional device in the process of technology development, where capability rapidly moved from the public sector to the private sector. They stated that the "Taiwan government's approach to the upgrading of technological capabilities within industry has been pursued using innovative institutional frameworks over the course of the three decades, 1965 to 1995. These frameworks have co-evolved with the industries they have fostered. The major sources for leverage have been training and engineering development: multinational investments and joint ventures; institutional support in the form of ITRI/ERSO; infrastructure such as the Hsinchu Science-based Industry Park."

The ITRI is the powerhouse behind Taiwan's entry into semiconductors. It is a source of information, manpower, and other advanced technology for the semiconductor industry. The ITRI has successfully built close working relationships with both private and governmental sectors. As a quasi-state-owned RandD institute, it conducts precompetitive research on projects sponsored by the Ministry of Economic Affairs, with a view to transferring the outcomes to the private sector nonexclusively. ITRI also conducts short- and medium-term research projects for the private sector, and is engaged with industry associations in the formation of various RandD collaborative consortia, designed to keep Taiwan firms abreast of world technological best practice. The ITRI's organizational goals in technology leverage and technology innovation, and against profit-seeking have earned the ITRI partner relationships with both government agents and private firms.[9]

[9] The mission of ITRI is technology transfer. It is arguably the most capable institution of its kind in the world in scanning the global technological horizon for developments of interest to Taiwan industry and executing the steps required to import the technology. Projects undertaken by ITRI were designed both to facilitate the creation of new industries, as in the case of semiconductors (but also fine chemicals, pharmaceuticals, optoelectronics, aerospace), and to upgrade existing industries. The semiconductor industry has certainly been ITRI's greatest industry creation success story.

4.4. The Role the ITRI Plays in Implementing Environmental Policy

The ITRI is not only the engine for developing new industries, it also plays a strategic role in implementing environmental policy in Taiwan's high-tech industries. There are three important factors that support this role. First, from the ITRI's perspective, clean technology and environmental management systems are attractive, innovative technologies on the technology horizon which will increase Taiwan's economic competitiveness in both high-tech and traditional industries. The ITRI, therefore, has devoted research toward implementation of ISO 14000, life-cycle analysis, eco-labeling, waste reduction technology and air pollution control technology, and so on, in Taiwan firms. Second, due to its technology capacity and social network relations with the industry, the ITRI was requested by the Taiwan EPA and IDB to explore greener technologies with firms to delineate the need for regulatory criteria. For example, the most recently announced air pollution emission standards incorporated one of the ITRI's projects to explore high-tech firms' technological capability in reducing air pollution emission. Based on its findings, the ITRI made suggestions to the Taiwan EPA to set the air pollution emission standards. Third, a cooperation and trust relationship has been developed between Taiwanese industries, especially high-tech industries, and the ITRI. Environmental management systems and clean technology promoted by the ITRI are likely to be accepted as tools to assist the competitiveness of firms, instead of being rejected as disruptive barriers to firms' profit seeking. As stated earlier, within Taiwan's high-tech industries, over 50 percent of firms have ISO 14000 certification. There are also ongoing waste reduction technologies and life-cycle analysis research projects on the part of the ITRI for Taiwanese firms. Clearly, with the help of ITRI, in terms of voluntary environmental technologies and environmental management systems, the rate of implementation is impressive for high-tech industries in HSIP.

5. DISCUSSION

5.1. Social Sphere: Environmentalism and Environmental NGOs

The development of environmentalism in Taiwan has considerably lagged behind the evolutionary thinking of environmentalism in Western societies (Hsiao 1994). Taiwan's environmental awareness was not awakened until the 1970s, and then was limited to a small group of people, mainly from the intellectual class. Not until the late 1980s did the general public's concern over environmental issues bloom.[10] In contrast with the West, Environmentalism in Taiwan from the 1970s to 1980s remained rather focused on local issues. Hsiao's (1994) study of Taiwan environmental movements from 1980 to 1991 revealed that, before the mid-1980s, the resources for mobilizing environmental movements and the emotional resolve for dealing with environmental problems were intertwined with the

[10] According to Hsiao's surveys in 1983 and 1986. In 1983, 70 percent of the public rated pollution as serious or very serious; by 1986, the figure had jumped to 88 percent. In 1983, 51 percent of the public thought pollution would get worse over the coming five years; by 1986, this had risen to 70 percent.

attempts of political liberalization movements.[11] Ecological rationalization, arguably, has thus followed an independent course, especially in its early stages.

The lift of 'Martial Law' as a key event in Taiwan's democratization was also a significant benchmark for Taiwan's environmental reforms. After the lift of 'Martial Law' in 1987, Taiwan's environmental NGOs increased the number and scope of issues addressed, from a focus on mainly environmental protection issues to more diverse and complex ecological issues. In the following period, environmental protection movements tended to be more organized and institutionalized. In Hsiao's survey of Taiwanese environmental NGOs (1997), resource sharing, and exchange and support relationships characterized the trend in the relationship between governmental environmental agents and environmental NGOs. However, tactics of competition, not just sharing, are also utilized within this symbiotic relationship.[12]

5.2. Political Sphere: Environmental Policy Transition

Examining environmental policy and regulatory frameworks, we may observe that there were increasing environmental legislation and regulations enacted in the past few years. The most significant institutional change was enacted by the Environmental Impact Assessment Act in 1994, in which consultation with academia, NGOs, and nongovernmental experts were required in establishing environmental regulations. Such a participatory policy making process is crucial in EMT. In addition, Taiwanese environmental policy has so far adopted few measures to economize ecology, such as air pollution and water pollution permits, and garbage-bag charges. Using economic incentives to achieve environmental protection goals was specifically stipulated in the 1998 the National Environmental Protection Plan. Lastly, beyond pollution control, preventive measures are implemented in the Taiwan environmental regulatory regime, as exemplified in the 1994 Environmental Impact Assessment Act.

5.3. Economic Sphere: Industrial Policy and Industrial Structure

A government must rely on cooperative relations when dealing with the industry on environmental issues. However, equally significant is the willingness of the industrial sector to accept the authority of the state, and the fact that government policies do make a difference to the relevant factors. In implementing rigid pollution control measures, governmental agents such as the IDB face a constant struggle to protect the vulnerable small- and medium-scale factories since they constitute a vital stream of economic growth for Taiwan.[13]

In a developmental state like Taiwan, the state plays a leading role in making blueprints for the industrial landscape. Details of economic development strategies were stipulated in the Economic Development Plan. In the 1986–1989 plan, pollution control first appeared in the

---

[11] The benchmark of the environmental movement is the removal of 'martial law' on July 15, 1987.

[12] An interview with a veteran environmental activist indicated that there is a transformation underway in the strategies of Taiwanese environmental movements from oppositional to participatory, from hostile to cooperative. It is also evident in an environmental protection movement in HSIP, where the slogan "Against Pollution, Not Against Business" is used.

[13] Information gained from interview with IDB staff.

agenda. The Taiwan government began to incorporate environmental considerations into economic development.[14] Under this guideline, several structural transformations are revealed. First, the latest competitive shift in industrial structure has moved from high-pollution industries to lower-pollution industries. The new high-tech industries promoted under the Statute for Upgrading Industry are considered relatively clean. Second, there are governmental economic development agents, other than the Taiwan EPA, involved in environmental protection efforts, such as the Industrial Development Bureau's "sustainable industries project," and the ITRI's facilitation of "eco-labeling," waste reduction, life-cycle analysis, and "ISO 14000" implementation.

The rationale of economizing ecology is embedded in the Taiwan government's motivation for incorporating environmental consideration into economic development. The increasing demand to manage the environmental crisis in Taiwan also creates immediate pressure for the Taiwan government to make an institutional response. Two considerations are also factored into the plan of promoting "eco-industries."[15] First, meeting the increasing international competitiveness demands for cleaner production. Second, anticipating future growth in the demand for environmental technology in Southeast Asia, which has a similar economic infrastructure and has experienced similar environmental degradation.

5.4. Embedded Autonomy and Environmental Policy in High-Tech Industries

There are increasing environmental concerns regarding Taiwanese high-tech industries. However, it is acknowledged that voluntary environmental management in high-tech industries, especially in HSIP, is comparatively better than in those traditional industries and those firms which are not situated in the HSIP.[16] The better environmental management systems in high-tech industries can be attributed to their intrinsic nature, which involve continuous technology innovation, making it possible for such companies to adopt the most advanced environmental technologies. Also, as export-oriented industries, intense international competition in high-tech industries makes firms vulnerable if environmental regulations are not met. In the case of Taiwan, the embedded social relationships between government agents, typically the ITRI, and high-tech firms directly facilitate the implementation of advanced clean technology and environmental management systems. This social network is not only manifested in the relationship between private and public sectors, but it is also manifested in the informal network organized by environmental and occupational safety professions in HISP firms. Unaffiliated firms share their environmental and occupational safety information informally, which creates an important learning environment for advancement in environmental management.

---

[14] Three methods are implemented for this purpose. First, an import-substitution industrial development strategy is being followed to create an indigenous environmental goods and services industry. Second, industry purchases of pollution control and abatement equipment are heavily subsidized. Third, the government is financing research into pollution prevention and providing industry with subsidized technical assistance in waste reduction/minimization.

[15] Information gained from interviews of IDB and ITRI staffs.

[16] HSIP was created by the government to provide an infrastructure to facilitate and accelerate the process of technological diffusion, and to localize the vertical division of labor among semiconductor firms.

CONCLUSION AND FUTURE RESEARCH

Two main questions are addressed in this study. First, the extent to which environmental reform is taking place in Taiwan, and what mechanisms bring about the reforms. Second, an examination of the usefulness of EMT concepts for evaluating and shaping environmental protection efforts in Taiwan.

A few arguments are developed here. The increasing strength of civil society is salient in Taiwan. Taiwan environmentalism is on the rise. With the development of democratization, the civil society sector has begun to gain a foothold to bargain with the government on environmental policy making, although the number of successful cases remains small. In terms of the advancement of environmental policy, environmental management techniques, the improvement in the measures of air and water quality, and the restructuring of industry, Taiwan shows promising signs of moving toward the path of ecological modernization. The ecological modernization process has taken place in Taiwan via the unique political economic structure of the embedded relationship between the state and business. Taiwan as a developmental state strategically uses its embedded social networks in promoting environmental management and technology as manifested in the role that the ITRI plays in environmental policy in high-tech industries. High-tech firms have increasingly adopted environmental practices consistent with ecological modernization hypotheses, such as ISO 14000, life-cycle analysis, the abandonment of CFCs, and other environmental management techniques. In short, in Taiwan, it appears that embedded social network relationships between the state and firms assist firms to take an ecological modernization path.

As for the Taiwan government's motives, findings in this chapter are consistent with what Rock (1996) has found in his research. The factors contributing to the Taiwan government's continuous efforts in environmental protection include the process of democratization, growing public awareness about environmental issues, international competitiveness, gaining first-mover advantage, and export opportunities for green technology to the less developed countries, as well as cost considerations.

To some degree we may find ecological modernization does cast some light on the dynamics of environmental reforms in Taiwan. However, the results of the dematerialization efforts are too early to judge. Although Taiwan's environmentalism is rather underdeveloped in comparison with the West, the prevalence of "sustainable development" discourse and the adoption of environmental management and other technical and policy innovations seem to suggest that it is catching up with its counterparts in the West. There are many environmental measures implemented, and environmental information is gathered and presented by the Taiwanese government. However, to what degree this is a consequence of symbolic functions rather than rational ecological choices remains unknown. Direct evidence of dematerialization and the results of environmental reform will not be realized for some years. Therefore, it would be fair to conclude that Taiwan environmental reform, in the measure of "process," does move toward ecological modernization. Nevertheless, in terms of the measure of dematerialization, this is not so conclusive.

EMT has been questioned about the applicability of explaining environmental reforms in less industrialized countries and newly industrializing countries. Taking on the challenge of EMT's applicability in less industrialized countries, Social scientists developed a research agenda intended to broaden EMT geographically.[17] Some are mainly concerned about the ways in which globalization processes might catalyze ecological modernization processes in less industrialized countries (Frijns et al. 2000). Others studied the applicability of EMT in newly industrialized countries (NICs), specifically focusing on the efforts of efficiency improvement, waste reduction, and cleaner production processes in pulp and paper industries among Southeast Asian countries in the late 1980s and early 1990s (Sonnenfeld 2000; Mol and Sonnenfeld 2000b). Sonnenfeld found that the greening efforts in pulp and paper industries in Southeast Asian countries partially fulfills EMT's hypothesis. However, there was no evidence of dematerialization in the cases he studied. Observing the uneven distribution of green practices, Sonnenfeld's research concluded that the dynamics of ecological modernization exist in large-scale, export-oriented modern sectors of NIC economies, but are less prevalent in small- and medium-size enterprises (SMEs).

Comparative studies in EMT across different developmental levels of countries test the global applicability of EMT. However, applying EMT in NICs thus far has returned no conclusive argument. The case of Vietnam challenges the force of globalization in modernizing ecological rationality in a less developed country. It implies that time and developmental stage matters, and the mechanisms of globalization of ecological modernization are hardly taking effect. That is, until economic development and/or political democracy have reached a certain level, ecological consideration in social and economic development has no visible position. In contrast, Sonnenfeld's study (2000) showed that globalization is effective in shaping environmental reform in NICs. Ecological modernization is happening in large-scale, export-oriented modern sectors of NIC economies. The study of Taiwanese environmental reform discussed in this chapter agrees with Sonnenfeld's conclusion that large-scale, export-oriented modern sectors like high-tech industries do show more evidence in taking an ecological modernization course. However, in the case of Taiwan high-tech, the leading role the state plays in promoting environmentally sound technology and management and the embedded autonomy state–society structure are compounded with the globalization force to lead high-tech industries to pursue ecological modernization.

In the globalizing world, with modern industrialized countries still providing the dominant models of economic development, models of ecological reform that are believed to be inappropriate for non-OECD countries might still be imposed upon these countries through a diversity of mechanisms.[18] Without real informed environmental knowledge and consciousness, the implementation of environmental regulation is not going to promise Taiwan a sustainable future.

In this chapter, social dynamics, environmental policy and industrial policy are carefully reviewed. Further data or information gathering on the business sector's rationalization of environmental reform is essential in answering the above questions.

In light of EMT, Taiwan's case brings out a challenge for future discussion. As discussed in the political economy literature, as Taiwan is an NIC, a developmental state, its economy is not completely independent from the state. Can we expect a future of 'ecological emancipation from the economic sphere'? In other words, how an ecological modernization

[17] See Mol and Sonnenfeld (2000b).

[18] E.g., the political mechanisms such as the international development programs of various nation-states and international actors (such as the World Bank and the IMF) or the emphasis put on the transfer of environmental technology from industrialized countries to developing countries in the UNCED declaration, the Montreal Protocol, and the elaboration on the Framework Convention on Climate Change.

process is going to proceed in a society where the functions of the political and economic spheres are much compounded remains to be explored.

ACKNOWLEDGMENTS

I would like to acknowledge Frederick Buttel for the initial inspiration for this chapter, Arthur Mol, David Sonnenfeld, Gert Spaargaren, Peter Ho, Michael Hernke and the journal's reviewer for many helpful comments, and Tze-Lun Lin and many anonymous others for assistance with data collection.

REFERENCES

Amsden, Alice H. (1985), "The State and Taiwan's Economic Development," in Peter S. Evans *et al.*, eds., *Bringing the State Back In*, Cambridge: Cambridge University Press.

Bahro, R. (1984), *From Red to Green*, London: Verso.

Beck, Ulrich (1992a), *The Risk Society*, London: Sage.

Beck, Ulrich (1992b), "From Industrial Society to the Risk Society: Questions of Survival, Social Structure and Ecological Enlightenment," *Theory, Culture and Society*, vol. 9, pp. 79–123.

Beck, U., Giddens, A. and Lash, S. (eds.) (1994), *Reflexive Modernization*, Polity, Cambridge.

Benton, Ted (1997), "Reflexive Modernization or Green Socialism?" Paper presented at the ISA RC-24 Conference on Social Theory and the Environment, Woudshcoten, The Netherlands.

Bluhdorn, Ingolfur (1997), "Ecological Modernization and Post-Ecologist Politics," Paper presented at the ISA RC-24 Conference on Social Theory and the Environment, Woudshcoten, Netherlands.

Bookchin, M. (1989), *Remaking Society*, Montreal: Black Rose Books.

Buttel, Frederick H. (1999), "Classical Theory and Contemporary Environmental Sociology: Some Reflections on the Antecedents and Prospects for Reflexive Modernisation Theories in the Study of Environment and Society," in G. Spaargaren, A.P. Mol and F. Buttel, eds., *Environment and Global Modernity*, London: Sage.

Byster, L. and Smith, T. (1999), High-Tech and Toxic. *Forum for Applied Research and Public Policy*, vol. 14, no. 1, pp. 69–75.

Castells, Manuel (1992), "Four Asian Tigers with a Dragon Head: A Comparative Analysis of the State, Economy, and Society in the Asian Pacific Rim," in Richard P. Appelbaum and Jeffery Henderson, eds., *States and Development in the Asian Pacific Rim*, Newbury Park: Sage Publications.

Catton, W.R., Jr. and Dunlap, R.E. (1980), "A New Sociological Paradigm for Post-Exuberant Sociology," *American Behavioral Scientist*, vol. 24, pp. 14–47.

Chiou, Chang-Tay (1995), *Environmental Regulatory Policy in Taiwan*, Taipei: Shu-Shin (in Chinese).

CEPD, (1999), *The R.O.C.'s Economic Development Plan*, Council for Economic Planning and Development, Executive Yuan, The Republic of China. (www.cepd.gov.tw).

Deyo, Frederic C. (1987), ed. *The Political Economy of the New Asian Industrialism*, Ithaca, NY, Cornell University Press.

Environmental Protection Administration (1999), *Environmental Policy Monthly*, February, Taipei, Taiwan.

Evans, Peter (1995), *Embedded Autonomy: States and Industrial Transformation*, New Jersey: Princeton University Press.

Evans, Peter (ed.) (1997), *State-Society Synergy: Government and Social Capital in Development*, California: University of California at Berkeley.

Frijns, J., Phung Thuy Phuong and A. P. J. Mol (2000), "Ecological Modernization Theory and Industrializing Economies. The Case of Vietnam" In: A.P.J. Mol and D.A. Sonnenfeld (eds), *Ecological Modernization Around the World: Perspective and Critical Debates*. London, Portland: Frank Cass. 257-292.

Giddens, A. (1998), *The Third Way*, Polity, Cambridge.

Grabher, G. (ed.) (1993), *The Embedded Firm, On the Socioeconomics of Industrial Networks*, London and New York: Routledge.

Hajer, Maaren A. (1995), T*he Politics of Environmental Discourse: Ecological Modernization and the Policy Process*, Oxford University Press.

Hsiao, Hsin-Huang Michael (1982), " Environmental Quality and Environmental Problems: A Public Perception and Attitudes" (in Chinese), in Chang-I Wen, Cheng-hung Liao, H.H. Michael Hsiao, and K.J. Chen, eds., *Social Change in Economic Development Processes*, pp. 29–81, Taipei: Community Development Research and Training Center.

Hsiao, Hsin-Huang Michael (1986), "New Environmental Paradigm and Social Change: An Analysis of Environmental Values in Taiwan" (in Chinese), *Journal of Sociology* (National Taiwan University), vol. 18, pp. 81–134.

Hsiao, Hsin-Huang Michael (1989), "The Rise of Environmental Consciousness in Taiwan," *Impact Assessment Bulletin*, vol. 8, no. 1–2, pp. 217–231.

Hsiao, Hsin-Huang Michael (1991), "Public Perception and Attitudes Toward Environmental Protection" (in Chinese), in *The Social Attitudes Survey of Taiwan: The Report on the Feb. 1991 Survey*, pp. 90–97, Taipei: ISSP, Academia Sinica.

Hsiao, Hsin-Huang Michael (1992), "Formation and Transformation of Taiwan's State–Business Relations: A Critical Analysis," *Bulletin of the Institute of the Ethnology, Academia Sinica*, vol. 74, Autumn, pp. 1–32.

Hsiao, Hsin-Huang Michael (1994), "Taiwan Environmental Protection Movement's Characters and Transition: 1981-1991." In: *Environmental Protection and Industrial Policy*, pp. 550-573, Taiwan Research Foundation ed. Taipei: Chien-Wei Press.

Taiwan Research Foundation ed. (1997), *Local Environmental Protest Movements in Taiwan: 1991-1996*, Taipei, Taiwan: Environmental Protection Administration (in Chinese).

Taiwan Research Foundation ed. (1997b), *A Symbiotic Relationship in Tension: The Cooperative Relationship between The Environmental Protection Administrative Agency and NGOs*, Taipei, Taiwan, Environmental Protection Administration (in Chinese).

Taiwan Research Foundation ed. (1999), "Environmental Movements in Taiwan," in Yok-Shiu F. Lee and Alvin Y. So, eds., *Asia's Environmental Movements: Comparative Perspectives*, New York and London: M.E. Sharpe.

Ho, Peter, and Richard L. Edmonds (2007), *Embedded Environmentalism: Opportunities and Constraints of a Social Movement in China*, London: Routledge

Hsiao, Hsin-Huang Michael, Millbrath L.W., and Weller, R.P. (1995), "Antecedents of an Environmental Movement in Taiwan," *Capitalism, Nature, Socialism*, vol. 6, no. 3, pp. 91–104.

Huang, J. T. (1994), *Environmental Law in Taiwan*, Taipei: Yue-Dan. (in Chinese).

Huber, J. (1989), "Social Movements," *Technological Forecasting and Social Change*, vol. 35, pp. 365–374.

Janicke et al. (1989), "Structural Change and Environmental Impact," *Environmental Monitoring and Assessment*, 12, 2:99-114.

Lin, Tze-Luen (1995), *Participation of Environmental Interest Groups in Taiwan*. Paper presented at the First Annual North American Taiwan Studies Conference, Yale University, New Haven, Connecticut, June 3–4.

Lovins, A.B. (1977), *Soft Energy Paths, Towards a Durable Peace*, New York: Harper and Row.

Mathews, John A. and Cho, Dong-Sung (2000), *Tiger Technology: The Creation of a Semiconductor Industry in East Asia,* Cambridge University Press.

Meadows, D.H. *et al.* (1972), *The Limits to Growth.* New York: Universe.

Mol, Arthur P.J. and Sonnenfeld, D. (2000a), "Ecological Modernization Around the World: An Introduction," *Environmental Politics*, vol. 9, no. 1, Spring, pp. 3–16.

— (eds.) (2000b), *Ecological Modernization Around the World: Perspectives and Critical Debates*, London: Frank Cass.

MOEA (1999), "Industrial Statistics," Ministry of Economic Affairs, The Republic of China. (http://www.moeaidb.gov.tw).

Paterson, W. (1991a) "Self-Regulation under Pressure: Environmental Protection Policy in the Chemical Industry and the Response of Sectoral Business Associations," in L.A. Martinelli, ed., *International Markets and Global Firms. A Comparative Study of Organized Business in the Chemical Industry*, London: SAGE.

Paterson, W. (1991b) "Regulatory Change and Environmental Protection in the British and German Chemical industry," *European Journal of Political Research*, vol. 19, 307-326.

Rock, Michael T. (1996), "Toward More Sustainable Development: The Environment and Industrial Policy in Taiwan," *Development Policy Review*, vol. 14, pp. 255–272.

Schnaiberg, A. (1980), *The Environment.* New York: Oxford University Press.

Schnaiberg, Allan, Weinberg, Adam, and Pellow, David N. (1999). "The Treadmill of Production and the Environmental State," Paper presented at the ISA/RC24 Conference on the Environmental State Under Pressure, Chicago, August 1999.

Simonis, U.E. (1989). "Ecological Modernization of Industrial Society: Three strategic Elements," *International Social Science Journal*, vol. 121, pp. 347–361.

Sonnenfeld, David A. (2000), "Contradictions of Ecological Modernisation: Pulp and Paper Manufacturing in South-East Asia", *Environmental Politics*, vol. 9, no. 1, Spring.

Spaargaren, Gert (1997), "The Ecological Modernization of Production and Consumption, Essays in Environmental Sociology," Dissertation, Wageningen: Department of Environmental Sociology WAU.

United Microelectronics Corporation (2000), Annual Report. Taiwan.

Wade, Robert (1990). *Governing the Market: Economic Theory and the Role of Government in East Asian Industrialization*, Princeton, NJ: Princeton University Press.

Weale, Albert (1992), *The New Politics of Pollution*, Manchester: Manchester Univ. Press.

World Commission on Environment and Development (1987), *Our Common Future*, Oxford: Oxford University Press.

Wang, C. S. (1998), "Economic Growth with Pollution: Taiwan's Environmental Experiences," Chapter 5:121-151 in Uday Desai, ed., *Ecological Policy and Politics in Developing Countries: Economic Growth, Democracy and Environment,* New York: SUNY Press".

Yang, Li-Fang and Lin, Tze-Luen (2000), "The State, High-tech Industries, and Environmental Reforms in Taiwan: An Ecological Modernization Perspective," Presented at The Sixth Annual Conference of the North American Taiwan StudiesAssociation, Harvard University, Massachusetts, June 16–19.

In: Leapfrogging Development in Emerging Asia
Editor: Peter Ho, pp. 151-172
ISBN: 978-1-60456-410-5

Chapter 8

ZERO LANDFILL, ZERO WASTE: THE GREENING OF INDUSTRY IN SINGAPORE

Josephine Chinying Lang
Nanyang Business School,
Nanyang Technological University, Singapore

LIST OF ACRONYMS

ACCA	The Association of Chartered Certified Accountants
BCA	Building and Construction Authority
CandD	Construction and demolition
EDB	Singapore Economic Development Board
EnnovaTe	Innovation in Environmental Technology
ENV	Ministry of Environment
EPA	Environmental Protection Agency
ETI	Environmental Test-Bedding Initiative
HCH	Hock Chuan Hong Waste Management Pte Ltd
IES	Innovation for Environmental Sustainability
JTC	Jurong Town Council
LTA	Land Transport Authority
NEA	National Environment Agency
NICs	Newly Industrializing Countries
NTU	Nanyang Technological University
NUS	National University of Singapore
PUB	Public Utilities Board
SCAL	Singapore Contractors Association Limited
SGP 2012	Singapore Green Plan 2012
SINERGY	Singapore Initiative in New Energy Technology
SMFs	Small- and medium-sized factories

1. Introduction

Singapore is a Southeast Asian island city-state of only 682 square kilometers and a population of about four million. In the 1970s and 1980s, its environmental management was focused largely on pollution control with stringent legislation on water, air, and waste disposal. In this sense, Singapore as one of the first-generation Newly Industrializing Countries (NICs) has shown successes where other NICs, such as Taiwan, Hong Kong, and South Korea faced much greater environmental challenges. Having effective pollution control measures may be reflected in Singapore's reputation for being clean and green. The hygienic environment is credited with enabling Singapore to quickly check the outbreak of the severe atypical respiratory syndrome (SARS) epidemic in 2003 and keep at bay the bird flu epidemic that impacted ten Asian countries in 2004.

Robust economic growth in the 1980s and 1990s saw Singapore's income rising rapidly to that of developed nations. Rising in tandem with its increasing consumption was waste generation. For example, the volume of solid waste increased approximately sixfold from 1,260 tons/day in 1970 to about 7190 tons /day in 2002. Such a trend, if left unchecked, would mean the need for a new incineration plant every five to seven years and a new landfill every 25-30 years (ENV, 2001). Thus, pollution prevention and environmental sustainability became the main foci in the 1990s. To meet these challenges, closing the solid waste loop through waste minimization, the adoption of green technologies, and recycling became a priority. An ambitious goal calling for "zero landfill and zero waste" was mooted to reduce as much as possible the incineration of waste by promoting the recycling of waste and the innovative use of recycled materials. The Singapore Green Plan 2012 (SGP 2012) proposed to increase the recycling rate to 60% by 2012, strive towards zero-landfill, and reduce the need to build new incineration plants from the current five to seven years to every 10 to 15 years or longer (ENV, 2002d). Public-private collaboration to develop new environmental technologies was also stressed.

This chapter consists of three sections. The first section provides an overview of solid waste management in Singapore, especially its key challenges arising from a shortage of land and escalating costs of incineration. The second section outlines how it has implemented a two-pronged greening strategy, i.e., development of a recycling industry and the initiation of strategic partnerships. Four case studies follow to illustrate the multiplayer collaborations and the facilitative roles played by various government authorities. Of particular interest is the role of the government in reconciling environmental sustainability and business profitability. The third section summarizes the progress made to date and discusses critical issues relating to environmental reporting in Singapore.

2. Solid Waste Management

The Ministry of the Environment (ENV) formed in 1972, through its statutory boards, namely, the National Environment Agency (NEA) and the Public Utilities Board (PUB), manages sewerage, drainage, and solid waste disposal; monitors air and water pollution, controls hazardous and toxic wastes; and provides environmental public health services as well as public education on waste minimization and recycling. In order to promote waste minimization aggressively, the Waste Minimization Unit (currently known as the Waste Minimization Section under National Environment Agency) was established in February 1992. A specific waste management hierarchy was emphasized: waste prevention and reduction was the highest priority followed by reuse, recycle, and recover in that order.

Environmental public policy in Singapore had largely been driven by various media-based environmental legislation and regulations such as the Clean Air Act and Regulations 1971/73 and the Water Pollution Control and Drainage Act 1975. These were consolidated in April 1999 as the Environmental Pollution Control Act. But there has been a significant shift in the government's management approach, moving from "top-down, command and control" approach towards greater industrial self-regulation and collaborative initiatives. The injection of competitive market forces into environmental management in Singapore has resulted in greater participation of industry players in promoting environmental sustainability. There are about 400 companies in the waste management and recycling industry in Singapore. Waste collection and removal are the responsibility of four private companies. NEA awarded each company a five-year contract to collect refuse and to take part in island-wide recycling (NEA, 2002-2003).

While domestic waste collection went private in 1999, ENV continues to control and operate the four incineration plants situated at Senoko, Ulu Pandan, Tuas, and Tuas South. During incineration, energy is captured and converted into electricity and ferrous metals are recovered to be recycled. Still, the role of the government in waste disposal will diminish significantly with the privatization of these incineration plants. In 2005, the Tuas and Tuas South incineration plants will be privatized. A fifth plant, owned and run by a private company, is expected to be completed in 2006. Once this fifth plant is fully operational, the Ulu Pandan Incineration Plant will be shut down. The only landfill, however, will not be privatized, as there is no market competition in this area (Lau, 2000).

There are two main categories of waste in Singapore: first, domestic and trade waste collected from residential premises, food centers and markets, schools and trade premises; and second, commercial and industrial waste collected from commercial and industrial premises, construction sites and shipyards (NEA, 2002a). The volume of waste increased steadily from 739.9 thousand tons in 1976 to 2625.6 thousand tons in 2002, an increase of three and a half times. Figure 1 shows the amount of waste disposed at landfills or burnt in incinerators for the last three decades. Total disposed refuse increased steadily from 1976 to the peak of 2.8 million tones in 1998 and stayed at that level for three subsequent years. However, a decline began in 2002, which continued into 2003. A decreasing volume of waste was sent to incineration plants and the solitary landfill. There were two possible reasons for the decline: effectiveness of recycling programs launched in 2002 and the lagged effect of the 2001 recession, the worst one encountered in Singapore since independence in 1965. Recycling did take on a new level of significance when the National Recycling Program was introduced in April 2001. The troughs found in the volumes of refuse sent to dumping grounds between 1976 and 2003 coincided closely with the peaks in volumes sent to incineration plants. Each matching trough and peak occurred in the year when an incineration plant became operational. Of the total waste sent to incineration plants or the landfill in 2003, approximately 42.7 percent was industrial and commercial waste (NEA, 2002e). Figure 2 shows the composition of total solid waste (i.e., waste disposed in landfill, burnt in incinerators, and recycled) in Singapore in 2003.

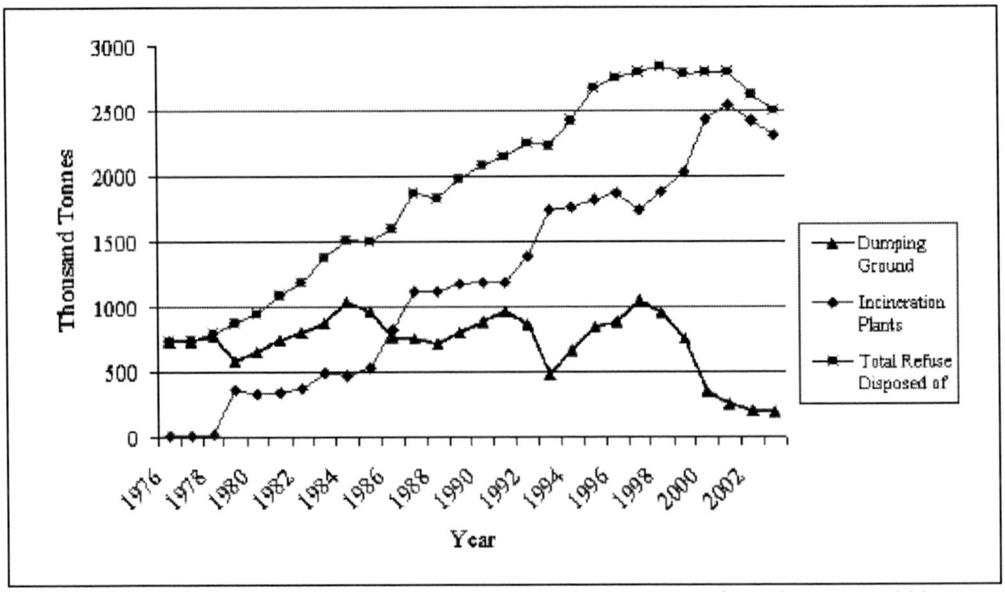

Sources: Ministry of Environment, 1998, Annual Report; Ministry of Environment 1999, Annual Report; Ministry of Environment, 2000, Annual Report; app.nea.gov.sg/cms/htdocs/article/asp?pid=401;
app.nea.gov.sg/cms/htdocs/article.asp?pid=1469.

Figure 1. Refuse disposed of at authorized disposal sites (1976-2002).

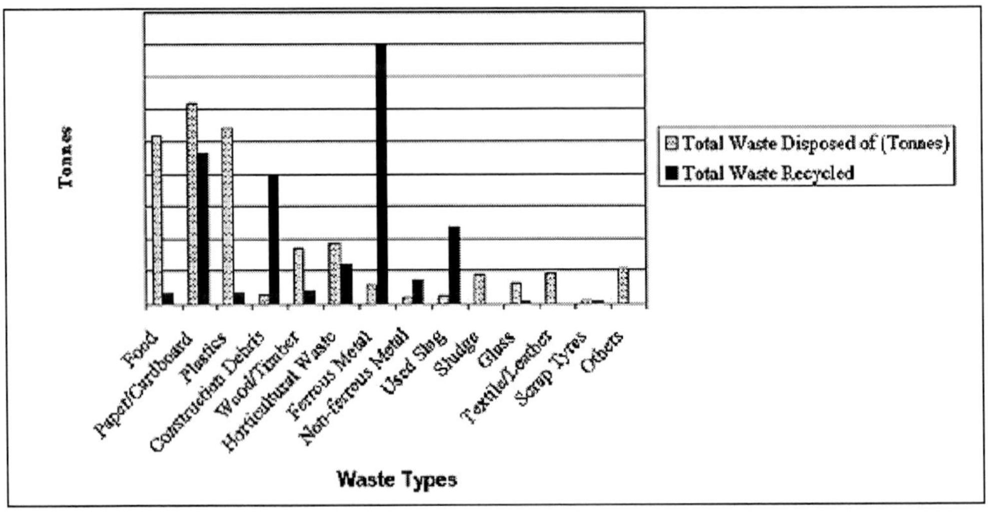

Source: NEA, (2004), app.nea.gov.sg/cms/htdocs/article.asp?pid=1469 accessed on July 14, 2004.

Figure 2. Waste disposed of and waste recycled, 2003.

Today, incineration is the main solid waste disposal method used in Singapore. In 2003, a total of 2.31 million tons of waste or about 92 % of the total refuse generated in Singapore were incinerated. During the incineration process, energy is captured and scrap metal recovered. In 2001, for example, 1159 million kWh of electricity were generated and 23,903 tons of scrap metal were recovered for recycling (NEA, 2004b; NEA, 2002a; ENV, 2002c).

While incineration has been very effective in processing large volumes of combustible waste, it is wasteful of nonrenewable resources (NEA, 2002a). First, although it helps to reduce the volume of combustible waste, there is still a need to landfill the noncombustible waste and incineration ash left over from the incineration process. Secondly, building waste disposal facilities can be very expensive. An incineration plant costs about S$500 million to S$1 billion to build (US$1=S$1.7); the Tuas South Incineration Plant, for example, cost about S$900 million.

Landfilling is not a better option either. Up to 1979, nearly all waste was disposed by dumping in five landfills situated on the main island. Upon reaching capacity, the five landfills were closed, the last one in 1999. A new landfill was constructed at Pulau Semakau where noncombustible waste such as concrete slabs, bulky waste materials, and incineration ashes is sent. This landfill, which costs about S$610 million to construct, is expected to reach its full capacity by 2030. Setting up another landfill would mean utilizing valuable land and financial resources that could be used for other purposes. Thus, reducing, reusing and recycling remain important solutions to waste management.

3. Towards the Greening of Industry

Singapore has a highly industrialized economy with strong focus on the manufacturing and service sectors. Foreign multinationals together with a few large domestic firms dominated the industrial scene in Singapore. Small- and medium-sized enterprises continue to play only a minor role in the country's economy. Agriculture and mining are not significant at all. Manufacturing, which accounted for 24.3% of GDP in 2002, remains the main engine of growth, with electronics taking the lead. The chemicals industry takes second place, followed by petroleum refining, which is the third largest in the world.

In terms of its environmental performance, Singapore took eighth place among 36 countries in the 2001 GIN-DEX survey using World Bank and International Energy Agency data on carbon emission, water pollution, commercial energy use, and industrial output. The index, the product of a cooperation among the Greening Industry Network, Chulalongkorn University, and the US-Asia Environmental Partnership, captured the synergy between economic and environmental improvement in each country surveyed (Mccoy, 2001).

Singapore's approach to the greening of its industry is two-pronged in orientation: first, resource conservation through a more intensive and coordinated effort at developing the recycling industry; and second, waste reduction and elimination at the source through product and process innovations as well as developing green technologies for industry through private-public collaborations. The next section will discuss efforts taken to accelerate and streamline the development of a recycling industry in Singapore.

Developing the Recycling Industry

Aggressive recycling efforts in 2002 resulted in a 6.3% decline in waste output from 2,802,200 tons in 2001 to 2,625,600 tons in 2002. The overall recycling rate increased from 44.4% in 2001 to 47% in 2003. Figure 3 presents the percentage of waste disposed of and waste recycled in 2003. While construction debris, ferrous metal, and used slag had the highest recycling rates of more than 90%, food, plastics, glass, and textile/leather had the lowest recycling rates of below 10 percent. Table 1 presents the types of waste recycled and the different kinds of recycled products produced in Singapore.

In order to encourage industrial waste recovery and recycling, waste disposal fees have been raised slowly to reduce the need to subsidize. For example in 2002, incinerating one ton of refuse would cost S$87, while the disposal fee charged was S$67 per ton. In May 2002, this fee was raised to S$77 per ton to encourage waste reduction, minimization, and recycling (Feedback Unit, 2002).

To speed up the learning and adoption of environmentally friendly practices NEA, JTC, Singapore Manufacturers' Federation, the Waste Management and Recycling Association of Singapore, and various representatives of private companies came together to jointly produce a "Guidebook on Waste Minimization for Industries." The objective of the Guidebook is the dissemination of recycling information to promote greater efficiency and effectiveness in using, reusing and recycling resources by providing practical advice on waste minimization. The Guidebook showcases model companies that have achieved significant cost savings by reducing waste generation.

Two strategies are particularly significant in accelerating the development of a viable recycling industry in Singapore: the establishment of two recycling parks and the introduction of aggressive recycling programs in various high-rise industrial estates.

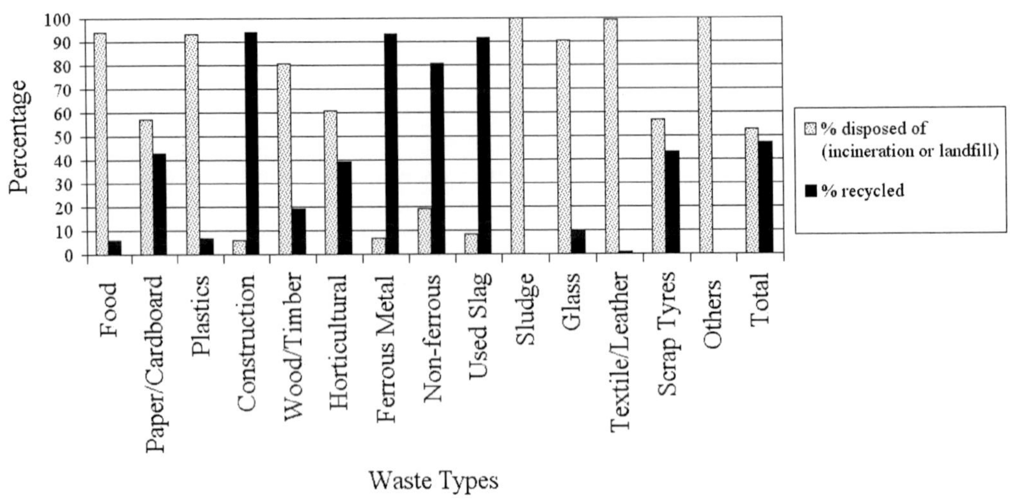

Source: NEA. 2004. Waste Statistics and Recycling Rates, app.nea.gov.sg/cms/htdocs/article.asp?pid=1469 accessed on July 14, 2004.

Figure 3. Percentage of waste disposed of and waste recycled in 2003.

Table 1. Recycling of Waste in Singapore

Types of waste	Recycled Products
Horticultural waste and wood wastes	Charcoal, compost, fertilizer
Soya bean wastes, spent grain and spent yeast	Animal feed
Used copper slag	Paving blocks and processed copper slag for ship blasting
Wood pallets, wooden crates, cases	Technical wood, flooring and building materials, new wood pallets and crates
Construction and demolition waste	Recovery of materials and manufacturing of precast concrete blocks from recycled concrete waste
Scrap metals	Iron rods
Scrap tires	Remanufactured solid industrial, trucks, and car tires, retreaded tires
Preconsumer plastics	Pellets
Waste concrete aggregates	Graded aggregates
Industrial waste	Recovery of materials
Electronics scrap	Recovery of precious metals

(a) Recycling Parks

NEA hopes to recycle 60% of all the waste generated in Singapore. Figure 4 presents the target rates of recycling for different waste types. The recycling industry took off in a big way only after the government took the initiative to create recycling parks offering low rentals. Two recycling parks have been created. First, the Sarimbum Recycling Park was created to attract the setting up of recycling facilities with low cost rental. Sited at a landfill closed in 1992, this is a 30-hectare site. About 2.5-hectare was leased to the first recycling plant to reuse wooden pallets and crates. A second recycling plant was established soon afterwards to convert wood chips, leaves, branches, and other trimmings into fertilizers or soil conditioners. Over the four subsequent years, seven other recycling plants were set up to produce rubber floorings from scrap tires and building materials from construction debris. Currently, twenty recycling plants are found on site (NEA, 2004b). Unlike existing construction and demolition recyclers that focus on low value-added utilization of waste the most recent addition, a Singapore-Australia 75:25 joint venture, will recover and process higher value-added secondary construction materials (ENV, 2003b).

The second recycling park, the Ecopark is a 19-hectare site situated beside the Tuas incinerator and other industrial estates. A site for high value-added recycling, the area clusters recycling companies together to leverage on economies of scale. This is where the Asia's first automated Material Recovery Facility, a S$7 million Singapore-Australia 60:40 joint venture, started its trial operations in Sept 2002, sorting recyclable household refuse such as used plastics, glass and metals for export overseas to be recycled. Recycling facilities for plastic and glass will be added as they become economically feasible with sufficient buildup in volume.

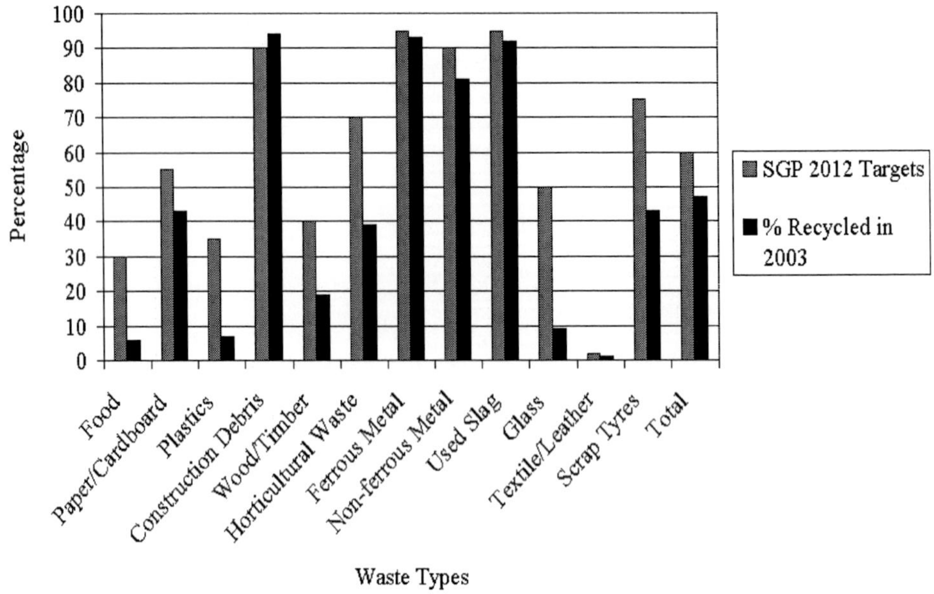

Source: Singapore Green Plan 2012.

Figure 4. Percentage of waste recycled in 2003 and the Singapore Green Plan 2012 targets.

(b) Recycling at Industrial Estates

Besides the setting up of recycling parks, the NEA is also actively promoting recycling among small- and medium-sized factories (SMFs) by collaborating with the Jurong Town Council (JTC). The involvement of JTC is critical as it is Singapore's leading developer and manager of industrial facilities and business parks. It manages 70% of industrial land area. With 38 industrial estates under its wings, it has direct access to more than 7,000 companies, 2,000 of which are SMFs situated in high-rise factory buildings.

SMFs in Singapore throw out 3,200 tons of solid waste every month that will take five Olympic-size swimming pools to hold. Yet, most SMFs do not have the critical volume to make recycling an economically viable alternative to incineration. In order to manage the whole waste chain more effectively and efficiently, common recycling programs are needed to coordinate waste separation and promote recycling.

In 2002, NEA began to collaborate with JTC and SembVisy Recycling to start a pilot recycling program at one of the industrial estates called the Ayer Rajah Industrial Estate, which has more than 150 SMFs. During the experimental period, about 75,000 kg of wood waste and 3,000 kg of paper, plastics, metal and glass were recovered for reuse or recycling each month in the one-year pilot program.

The success of this project led to the formation of a Joint Committee on Waste Minimization between NEA and JTC. The objective of the Joint Committee was to introduce formal recycling programs to 40% of JTC industrial estates over a three-year period. This began with a recycling program in November 2003 at one of JTC's 21 high-rise industrial estates called the Kallang Basin Industrial Estate where more than 200 factories are situated. With this program, segregation of waste and recycling have now been made a standard feature in waste collection contracts with all JTC multistory or flatted factories (ENV, 2003a).

Successful recycling programs require the support of new environmental technologies to transform waste into usable secondary materials that can compete successfully with conventional materials in the marketplace. Such recycled secondary materials must therefore possess qualities that are at least equivalent to, if not better than, conventional materials and yet be priced competitively. This is where the second prong of the environmental strategy, building of strategic partnerships, kicks in.

Strategic Partnerships for Waste Reduction

Strategic partnerships have been found to be useful in accelerating the development and adoption of green technologies. University research centers are important sources of technology for industry. Fostering strong linkages between university and industry helps in the exchange of technological knowledge. In fact, a domestic study comparing the various means of technology transfer found that the higher the industry commitment to participate in technology transfer through joint RandD projects, the more successful technology transfer practices are. The study also highlighted the crucial role played by the government in ensuring successful linkups by providing the funding and resources needed (Lee and Win, 2004).

A S$20 million Innovation for Environmental Sustainability Fund (IES Fund) was set up by the government to help finance the development of innovative environmental technologies through public-private collaborations. To enhance participation, NEA collaborates with town councils, community development councils, public waste collectors, recycling companies and research institutions. In February 2003, NEA launched the Singapore Environment Institute (SEI) to provide environmental management training and skills development and to facilitate knowledge sharing in sustainable development.

Besides providing financial assistance, the government also makes it easy for companies with environmental projects to gain access to other resources with the launch of the Environmental Test-Bedding Initiative (ETI) in August 2003. This initiative, a collaborative effort among NEA, Economic Development Board (EDB), and Public Utilities Board (PUB), provides a one-stop platform for companies to obtain financial advice, infrastructure support, technical expertise, manpower, and regulatory supports. The ETI manages two programs, the Singapore Initiative in New Energy Technology (SINERGY) for advancing innovative alternative energy technologies and the Innovation in Environmental Technology (EnnovaTe) for promoting innovations and commercialization of new environmental products and technologies for water treatment, waste management, and pollution control.

Two environment technology projects took off under the wings of these two programs. Under SINERGY is an alternative energy project carried out by a consortium, comprising the Development Resources Pte Ltd, IdaTech LLC, Housing Development Board, Nanyang Technological University (NTU) and NEA, to test a fuel cell system that supplies electrical power for lighting in a multistory car park. Under EnnovaTe is a water treatment project, a partnership between Matrix Mambrane and PUB, to test membranes for the treatment of seawater and wastewater the success of which would mean the availability of an alternative source of water, a greater adherence to the required environmental standards for industrial wastewater discharge, and cost savings (EDB, 2004).

To demonstrate strategic environmental tie-ups among industry players, government agencies, and research institutions in Singapore, the case studies below deal with environmental reforms in the following industrial sectors: steel, construction, waste incineration, and food retail.

Case 1. The Steel Industry

Incorporated in the 1960s as the National Iron and Steel Mills, NatSteel has played an important role in the infrastructural and residential development in Singapore. In the 1980s, it broadened its operations by expanding into steel fabrication and construction-related products. Today, NatSteel has operations in some 12 countries and boasts of a turnover of S$1,544 million in 2002.

One common problem confronting steel mills is the disposal of waste slag, a nonmetallic material produced as a result of the steel making process that poses great economic and environmental costs. The greening of NatSteel is related to its ongoing efforts at transforming waste slag and ladle furnace slag into usable steel slag aggregate and asphalt mix respectively.

At NatSteel, after two years of RandD collaboration with the public sector, a way was found to convert waste slag into steel slag aggregate that can be used in road construction. With proper pretreatment, this by-product now not only meets stringent US EPA leaching tests but even outperforms natural rock, the conventional material used as road surfacing aggregate. Its superior quality includes better physical properties, binding properties, stronger interlocking structures, better skid resistance quality, stronger mechanical strength to withstand wear and tear, and better color retention. This translates into cost savings, as roads built with steel slag aggregate would require less maintenance and repair.

There is zero waste in the production of steel slag aggregate. During the processing stage, metals are recovered for reprocessing while slag dust is put into good use in the asphalt mix for road construction and resurfacing works. In Singapore, steel slag aggregate is now used widely in heavy traffic intersections, slip roads, major expressways, bus lanes and airport taxiways (Natsteel, 2003).

The secondary refining of molten steel results in another by-product called ladle furnace slag that heretofore has been disposed off by dumping at the Pulau Semakau Landfill. On average, about 24,000 tons of ladle furnace slag are produced in Singapore every year, enough to fill up three 1.7 meter high soccer fields at the cost of S$1.1 million. A four-year joint research effort by NatSteel and NTU has now successfully recycled ladle furnace slag into modified asphalt mix. The modified asphalt, when used in road construction, has been observed to have stable and high anti-stripping properties that make roads very durable. A site test, set up in collaboration with NTU, the Land Transport Authority (LTA) and the NEA, and partially funded by the IES Fund, is now underway to study the environmental impact of using modified asphalt mix (NTU, 2003). If successful, it will mean substantial savings from having to import granite, as Singapore has no local source of granite.

The green culture has seen NatSteel go into the environmental business itself through one of its subsidiaries called NatSteel EnviroTech (previously known as NatSteel Abrasives). NatSteel EnviroTech produces blasting abrasives from copper and steel slag for sale to shipyards, construction industries, and fabrication engineering industries within the country or around the region. It also recycles waste slag into premium roadstone materials, provides air quality testing services, treats and recycles both domestic and industrial wastewater into useful environmentally friendly products.

Case 2. The Construction Industry

The construction industry has played a significant role in transforming Singapore into a modern city. The challenge faced by the industry today is not only to achieve optimal use of land through intensive development, but also to strive for better environmental performance. This industry is both materials and energy-intensive and its greening focuses primarily on reducing its consumption of materials and energy.

Inefficient construction methods and resource utilization may account partially for the generation of construction waste. This can be prevented with conscious incorporation of environmental concerns during the building design stage and a more stringent resource management during the construction stage. Unfortunately, many studies have found that the environmental performance of the construction industry in Singapore leaves much to be desired (Ofori, 2000; Ofori, Briffett, Gang, and Ranashinghe, 2000; Tan, Ofori, and Briffett, 1999). The industry has a low level of awareness of environmental auditing and a general reluctance to take action other than those needed to meet regulatory requirements. It is not ready for systematic environmental management. Instead of using metal forms, permanent, or system formwork, and prefabricated elements, building contractors continue to use timber formwork, which contributes to deforestation. In addition, inefficient use of materials that resulted in high levels of wastage was noted even for companies that have implemented ISO9002. Clients were found to be even less interested in environmental protection and energy conservation than contractors.

In 2002, about 406,600 tons of construction debris were generated. Ninety percent of the waste was recycled while the remaining ten percent was disposed of at the offshore landfill. The high percentage of recycling, however, does not imply the end of the waste disposal problem for the construction industry. There remains a sizable volume of CandD waste sent for landfilling everyday, that is, about 170 tons in weigh and 340 cu meters in volume. In volume, CandD waste represents the second largest source of solid waste dumped at Pulau Semakau Landfill (BCA, 2002).

Yet, on a more positive note, the lack of environmental awareness in the construction industry has begun to change with more fervent involvement of NEA and Building and Construction Authority (BCA), emphasizing three key strategies for enhancing sustainability; first, recycling of construction and demolition (CandD) waste, second, minimizing waste through adoption of best practices in the industry, and, third, meeting ISO 14000 series of standards.

To support recycling of CandD waste, the NEA oversaw the setting up of several recycling facilities for concrete waste, general construction waste, and demolition waste. Currently, three recycling plants are in operation, converting CandD waste into low value added secondary aggregates for further processing into nonstructural concrete products for use in new building or as hardcore materials for temporary road access in construction sites. For example, the SembEnviro Alex Fraser Construction and Demolition Materials Recovery Facility plant that opened in September 2003 can transform CandD waste into higher value-added products and materials.

To ensure the long-term success of CandD waste recycling, the NEA also promotes the use of recycled materials in the construction industry. It encourages collaborations among recycling and construction companies and research institutions to explore the innovative use of recycled materials as substitutes for conventional construction materials and to examine the performance of these recycled building materials and products.

Pilot projects, involving NEA, PUB, LTA, NTU, and Hock Chuan Hong Waste Management Pte Ltd (HCH), have been created to test the suitability of precast concrete products such as precast concrete drains and recycled road kerb at some experimental sites. HCH had previously succeeded in manufacturing pre-cast concrete drain slabs from recycled aggregates. Meanwhile, another joint project among BCA, the National University of Singapore (NUS), RDC Concrete, WR Grace, and Eng Seng Construction, is conducting research into the use of waste to produce self-compacting concrete, a new construction material.

Under BCA, an experimental project is also underway to test the performance of concrete cubes with different mixes, compressive strength, and water absorption. For example, concrete mix produced with normal and recycled aggregates has been compared with regard to its workability and characteristic strength. Suitable concrete mix has been identified and used to produce precast products such as road kerb and channel drain. In its effort to encourage broader recycling efforts, BCA has also permitted the use of copper slag, waste generated by the ship repair industry, as partial replacement of fine aggregates (BCA, 2004).

Besides CandD waste, the construction industry also generates large volumes of soft marine clay from excavation works carried out at construction sites such as road and underground subway construction sites. Using IES Fund, the NEA is supporting a pilot project to explore the possibility of converting clay into a usable form to substitute for sand and earth in foundation engineering works. If this is proven to be technologically and commercially viable, land hungry Singapore will enjoy large cost savings as the need to import sand and earth for reclamation works will be reduced (BCA, 2002).

The construction industry is also trying to minimize waste by preventing pollution at itssource. One strategy is wider adoption of pre-fabricated concrete components, the production of which generate less solid waste, reduce noise and air pollution, and consume less energy compared to the traditional cast *in situ* method. Another strategy is to minimize waste generation at construction sites through dissemination of best practices. NEA, collaborating with BCA, the Singapore Contractors Association Limited (SCAL) and NUS, has launched several waste minimization projects. One of these benchmarks various waste levels and then identifies ways to eliminate the generation of waste. Knowledge and experience acquired from these joint projects are then disseminated widely across the construction industry through the Code of Practice for Demolition. The code provides guidelines related to the dismantling of reusable components of a building such as doors, windows, and cables, and the separation of waste for recycling.

In another project, seven local companies collaborated in developing an integrated concrete manufacturing and distribution system. This joint effort, named Project Dolphin, integrates the whole supply chain of raw materials for concrete manufacturing by containerizing inputs such as sand and aggregates at the source, essentially ensuring pollution-free environment by placing the entire production and delivery of ready mix concrete within an enclosed space. This integrated system enhances the overall environmental performance of the industry by reducing wastage from spillage due to multiple handling or dropping of aggregates onto public roads, and minimizing dust pollution by eliminating the use of tipper trucks and wheel loaders. This new system is expected to cut down the amount of land required for concrete production by 70%, and eliminate 50% of the trips made in transporting raw materials. In total, it confers a saving of $2.1 million (NEA, 2002d).

BCA plays another crucial role in promoting environmental sustainability in the construction industry by administering the ISO 14000 (Environmental Management System) Certification Scheme. Introduced in Singapore in 1996, the ISO 14000 series is a set of international standards on environmental management that address issues such as environmental management systems, environmental performance evaluation, environmental labeling, and life-cycle assessment. For example, ISO 14001 and 14004 are standards related to environmental management systems (EMS), while ISO14010, 14011, and 14012 are related to environmental auditing standards. The ISO 14001 standard on EMS is especially popular in Singapore as it provides a framework for achieving productivity gain and global competitiveness (NEA, 2002c). A financial scheme, the Local Enterprise Technical Assistance Scheme, has been put in place to aid small and medium enterprises to gain ISO 14001 certification. By December 2003, about 473 companies in Singapore had been awarded the ISO 14001 certification, among them the Tuas Marine Transfer Station/Semakau Landfill (Spring Singapore, 2004).

BCA encourages construction firms, developers and consultants, suppliers of building materials to strive for certification. It is compulsory for BCA-registered contractors in the top 2 grades (A1 and A2), who carry out public sector building and civil engineering works, to implement environmental management systems that meet ISO 14000 standards by July 2004. By November 2003, more than 60% of the 70 larger construction companies had already obtained certification. These forward-looking contractors will take a proactive role in monitoring their material wastage levels for concrete and steel, and consumption of water and energy.

Certified contractors are expected to monitor their waste generation and energy consumption, review construction methods or work processes to reduce waste, and when it is economically feasible, to adopt environment friendly technologies. Neo Corporation was the first construction company in Singapore to receive the ISO 14001 certification in 1998. The first developer to be awarded ISO 14001 certification was City Developments Limited, one of the major developers in Singapore.

Its environmental strategies include instilling a green culture in developing property projects, reviewing the environmental impacts at all stages of the development process, and soliciting the involvement of employees, consultants and contractors (BCA, 2003).

In sum, the construction industry is gradually greening through the joint efforts of various organizations to recycle CandD wastes, develop and use building materials made from recycled aggregates, and prevent pollution at source by advancing green technologies and achieving ISO 14000 standards. To date, significant progress has been made. Figure 5 shows that the sharp increase in construction debris in 2001 was closely matched by a rise in volume of recycled and a concurrent decline in construction debris disposed off by incineration or dumping. In 2003, the 94% recycling rate of construction debris even exceeded the SGP2012 target of 90% (Figure 4).

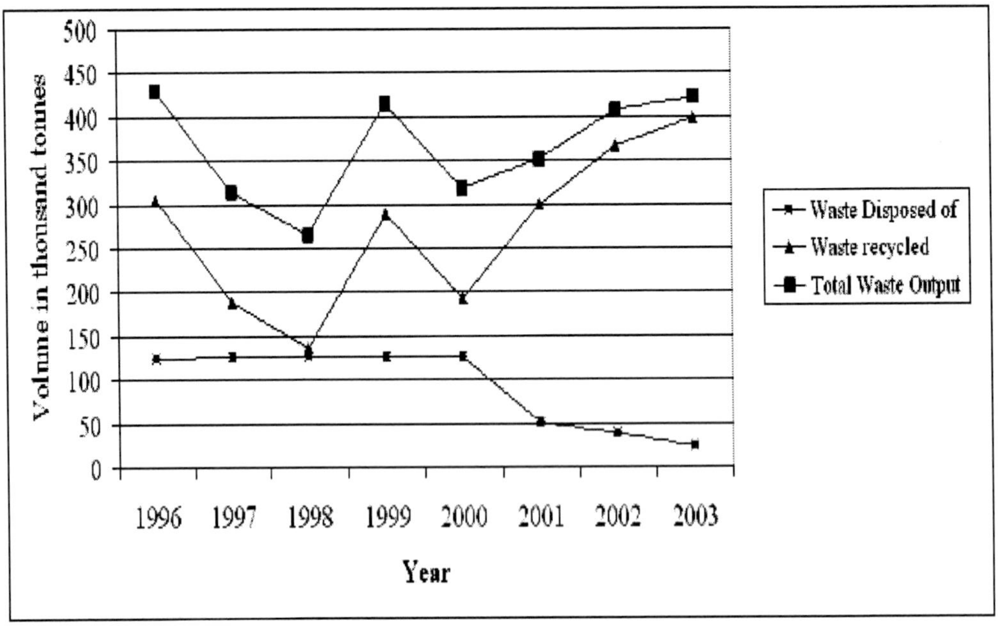

Source: NEA (2004). app.nea.gov.sg/cms/htdocs/article.asp?pid=1469 accessed on July 14, 2004.

Figure 5. Construction debris disposed of and recycled, 1996-2003.

Case 3. The Waste Incineration Industry

The third case-study reviews recent developments in the greening of the waste incineration industry in Singapore. Waste incineration generates waste too. Incineration ash is the waste left after incineration. On average, about 6,800 tones of waste are burnt everyday by the four incineration plants in Singapore, which produce about 1,600 tones of incineration bottom ash that have to be disposed of. There are two types of incineration ash; namely, bottom ash and fly ash. Bottom ash is the residue collected off the incineration grates, an inert material comprising of mostly glass, ceramic, rocks, and silica. Fly ash is the finer residue obtained downstream of the furnace. It is more hazardous than bottom ash as it contains easily leachable heavy metals and has higher lime content (Bai and Sutanto, 2002). To reduce the volume of incineration ash sent to the landfill, researchers in Singapore are exploring innovative methods to use incineration ashes for road construction, lightweight concrete products, soft soil stabilization, or for land reclamation (Show, Tay and Goh, 2003).

The suitability of using fly ash for road construction is still highly questionable, as environmental studies have shown that it may lead to leaching, where heavy metals and acid seep into the ground and pollute the groundwater. Much more research is needed to find ways to stabilize fly ashes that waste incineration produces. Untreated bottom ash can cause the destruction of road due to expansion and cracking. The swelling potential of bottom ash must therefore be tested prior to its use in road construction. Bottom ash must be treated to enhance its mechanical features and to reduce its potential chemical problems.

In May 2002, a strategic partnership among NEA, PUB, LTA, and Hanson Pacific Private Limited, an international building materials company with expertise in processing ash for road building in Europe, was formed to explore the feasibility of using bottom ash for road construction. By weight, the bottom ash represents 65% of the total waste disposed of at the Pulau Semakau Landfill; by volume, it represents 50%. At a S$77 disposal cost for each ton, this translates into a total sum of S$45 million a year for 1600 tons of bottom ash daily.

The S$74,000 project was partially financed by the IES Fund and carried out in two stages. In stage one, bottom ash was processed and screened to remove ferrous metals such as lead, nonferrous metals, and other impurities. It was then left to age for about eight to ten weeks. Laboratory tests were done to examine the suitability of its properties for road construction. Processed bottom ash was found to meet LTA's specifications for road-based materials. Leachate tests showed that the levels of heavy metals and other contaminants were kept within limits set by the United States Environmental Protection Agency, and its dioxins levels met the "Guidelines Values of Dioxin Levels in Soil" in Germany, the Netherlands, and Sweden.

In stage two, an experimental road 150 m long was built in February 2002 using 1,200 tons of processed IBA. To facilitate the collection of groundwater for testing, six groundwater standpipes were installed near the road. This road was opened to traffic in May the same year. Heavy vehicles ply this road frequently. After six months of monitoring, no cracks or localized sinking were found. This experiment showed that bottom ash could potentially be used for road construction as no adverse effects on groundwater quality and storm water runoff was observed. This experimental road also afforded a more comfortable ride than conventional roads based on the International Roughness Index used to assess road quality worldwide.

Armed with these results, NEA is moving to increase the use of bottom ash in road construction, holding talks with several companies to build a plant to process it for this purpose. It has been estimated that if recycled ash is used for road construction in Singapore, it would reduce 30% to 40% the amount of bottom ash sent to the Pulau Semakau Landfill (Kaur, 2003; ENV, 2002b; NEA, 2003). It would also reduce the need to use traditional road construction materials such as natural gravel and crushed rock, which are nonrenewable resources extracted from the nature. The NEA is also investigating the use of incineration ash for land reclamation.

Case 4. The Food Retail Industry

The final case-study deals with Singapore's efforts in the environmental reform of the food retail industry. In 2002, Singapore generated about 494,700 tons of food waste of which only 6% were recycled while 94% were incinerated. The food retail industry is fragmented with a multitude of players; 2,294 supermarkets, provision shops, mini-markets, and more than 11,000 market stalls operating in 150 locations spreading across the island (Canadian High Commission, 2004a). The food service industry is equally fragmented with 3,210 hotels, restaurants, fast food outlets and cafes, and more than 6,500 food stalls operating in hawker centers (Canadian High Commission, Singapore, 2004b). Many food waste generating sources are widely dispersed, which may account for the dismal recycling rate of 6%.

But why even incinerate? Landfilling of biodegradable organic material would seem appropriate if one assumes that this waste material decomposes and returns to nature. Unfortunately, this is not the case. Organic wastes, mixed with other types of wastes, do not return to their original state in nature. Landfills do not have the necessary conditions to allow for the completion of biological decomposition reactions to take place. Instead, landfills are "prison cells" for wastes. Thus, mixed wastes are deposited in landfills permanently separated from the environment by impermeable layers and marine clay to prevent contamination (Fehr,

Calcado, and Romão, 2002). However, incineration of food waste is not the most desirable as materials are destroyed. To close the food waste loop in an environmental way would require processes, technologies and management methods that eliminate the need to landfill or incinerate biodegradable organic waste.

SGP has set a goal of increasing the recycling rate of food to 30% by 2012 although so much food waste is generated everyday that must be treated before it can be discharged as a constituent of wastewater into the sewage system. The food waste in domestic wastewater is treated in a conventional activated sludge system in which are diverse microorganisms that give rise to a multitude of metabolic reactions leading to the degrading of organic matter.

Biological waste treatment plants used in restaurants, catering centers, food processing factories that process restaurant fats, oils and grease frequently stall due to the variability in the metabolic activities of microorganisms. Restarting these plants take time. The Productivity and Standards Board (predecessor of Spring Singapore) in collaboration with NUS and in cooperation with the Singapore Inflight Catering Center have found ways to reduce breakdowns and startup times. By reducing the biochemical oxygen demand (BOD) of the microorganisms, the efficiency of such treatment plants can be improved. By carefully choosing an optimum consortium of microorganisms, researchers have reduced the startup times of these plants (Fong and Tan, 2000).

Another similar project involving JTC and NTU, and largely funded by The Enterprise Challenge (TEC), was undertaken to implement biological treatment of industrial food waste at a JTC high-rise factory complex with more than 200 food processing companies. (TEC is a S$29m fund established to provide financial support and test beds for innovations that have potential in improving public service delivery.) The main aim of this project was to reduce the BOD level of industrial food processing effluents to levels permitted under the Sewerage and Drainage (Trade Effluent) Regulations 1999. The success of this project would mean an improvement in the efficiency of food waste treatment, less demand on the municipal wastewater system, more efficient use of land with the elimination for individual waste treatment systems, more economical waste treatment with the sharing of a treatment facility for high-rise food factories, and lower business cost with lower tariffs paid for pollutive trade effluents (ENV, 2002a).

NTU researchers have also been able to speed up conventional composting by using a mixture of sewage sludge and solid food waste to stabilize the organic waste, reduce smell and inactivate disease-causing microorganism. It is not easy for microorganisms to utilize the contents of sewage sludge technically because there is too much nitrogen relative to carbon in the mixture. Also, the viscous nature of sludge means that it is not well aerated whereas oxygen is needed for effective composting. NTU researchers, funded by PUB, have found that mixing sewage sludge with solid food waste lowers the amount of nitrogen relative to carbon and renders the texture of the mixture less viscous, so more oxygen is transferred into the bulk of the compost compared to sewage sludge alone.

The final product is a gray powder with only 5% moisture content. Usually, to accelerate composting of organic matter, finished compost, not cultures of microorganisms, is used. However, a starter culture made up purely of microorganisms has been found to do better because there is more control of desired metabolic processes with lower risk of disease-causing bacteria being found in the final product of biodegradation. By identifying a selected strain of a germ (Bacillus thermoamylovorans) that does not form spores, the researchers have also been able to prevent the growth of spores during biodegradation, which can cause allergies in human beings (Wang, Olena, Tay, Ivanov, and Tay, 2003).

A public-private collaboration in June 2003 saw a private firm, CPG Corp, install a 100 kg capacity composting machine costing S$350,000 at a site in the town of Tanjong Pagar where there are 160 food stalls which generate, on average, 1.5 tons of food waste daily. The machine now converts oily food scraps and innards into dry odorless compost in a process that takes about three to four hours. Compost produced is used to grow vegetables in a rooftop greenhouse nearby.

Initial reactions from the hawkers have been positive. Most cited a vast improvement in the hygiene of the place. For example, there were fewere pests such as crows, cockroaches and rodents while bad odors were eliminated as well. Others cited the economic benefits of 20% cost savings as the onsite composting machine eliminated the need to transport food waste to incineration plants. Once the commercial viability of this composting machine is proven, it will mean an increase in the food waste recycling rate. If food waste generated by all the wet markets and food centers in Singapore is converted into compost, the recycling rate of food waste rises very quickly to 19% (Teh, 2004), a large step closer to achieving the national goal of 30%.

Meanwhile, NEA and NTU are collaborating to develop an integrated food waste bioconversion system that will produce biogas and fertilizer from food waste. The generated biogas will be used to generate energy for recycling or composting while the fertilizer can be used for agriculture or landscaping.

4. Discussion and Conclusion

A shortage of land in Singapore for landfills and the escalating cost of incineration plants impelled the government to promote a recycling industry and form strategic partnerships among business firms, government agencies, and research institutions to develop innovative green technologies.

The case studies above illustrate the importance of strategic partnerships in the greening of industry in Singapore. Government agencies such as NEA, PUB, EDB, and BCA play an active role in building bridges between university research centers and industry players, initiating joint research projects and providing the necessary funding and regulatory support. Such collaborations not only enable more effective transfer of technology, but also ensure the economic viability of environmental projects. The imprimatur of government agencies in Singapore helps promote greater industry acceptance of recycled products and green best practices. Government involvement and financial support encourage business firms to invest in the exploration and testing of recycled aggregates as substitutes for conventional raw materials.

Energy inefficient and outdated technology and equipment incur not only high operating costs because they require more energy to operate but also high environmental costs because of pollutants emitted into the environment. Yet, the cost of acquiring energy-efficient technology may deter companies from making the necessary investment. To overcome this dilemma, two special incentive schemes are provided for under the Income Tax Act such that qualifying energy efficient or energy-saving equipment and pollution control equipment can be written off in one year (NEA, 2002b). First, a one-year accelerated depreciation allowance for energy-efficient equipment and technology encourages companies to replace energy inefficient technologies with energy efficient and environmentally friendly ones. Second, a one-year accelerated depreciation allowance for highly efficient pollution control equipment is designed to encourage investment in pollution control technologies. Under both these schemes, companies are allowed to depreciate 100% of the capital expenditure incurred.

To raise environmental awareness among domestic companies and encourage a proactive environmental management approach, the Singapore Environmental Council launched the Singapore Environmental Award in 1997. This award recognizes significant and innovative initiatives taken to reduce negative environmental impacts of production processes and products. Innovative initiatives may include attempts to conserve energy, enhance recyclability of products or packaging, reduce generation of toxic or hazardous wastes, and adopt clean production technologies. Selection criteria for the award include actions that go beyond compliance with environmental standards, rules and regulations; actions taken to address the cause rather than effects of pollution; the level of top management commitment and involvement; applicability of innovative initiatives to other companies; and finally, the environmental and commercial viability of innovations (Singapore Environment Council, 2004).

On the World Environment Day in 2004 another environmental award, the inaugural Singapore Green Plan 2012 Awards, was given. Formerly the Green Leaf Award, it gives recognition to excellent contributions by individuals, organizations and companies to protect the environment, especially significant achievements in enhancing environmental sustainability and attaining specific targets set out in the SGP. This Award is designed to encourage companies to go beyond fulfilling regulatory requirements in their environmental efforts by contributing to sound environmental practices and sustaining positive environmental impacts. Two companies won it in 2004, namely, STMicroelectronics Pte Ltd and HSBC. STMicroelectornics, a semiconductor company that had devoted 2% of its capital investment since 1995 to reduce the environmental impact of its operations, was recognized for its best practices in energy efficiency, waste and water recycling, and companywide proenvironment culture, while HSBC was recognized for its excellent efforts in nature conservation and environmental education as well as a pro-environment company culture (NEA, 2004a).

Another step in the greening of industry entails governmental regulatory pressure in the form of environmental self-reporting, or the regular public disclosure of environmental information such as waste volumes, pollution emissions, and pollution prevention efforts. Two surveys of the annual reports of public listed companies in Singapore found that voluntary disclosure was inadequate. At best, there was minimal disclosure on environmental policy, environmental benefits of products or processes, and announcements of ISO 14001 certification. At worst, information given was either uninformative or irrelevant. In general, quantitative data for tracking environmental progress were lacking, so were environmental targets or audits (ACCA, 2002a; Perry and Teng, 1999).

Possible reasons for such dismal environmental reporting include a lack of environmental awareness, perceived benefits and government pressure. Similarly the state of environmental reporting by the private sector in Hong Kong is dismal despite the fact that an environmental reporting initiative for all government departments, bureaux, and government-owned organizations had been launched way back in 1998. (ACCA, 2002b).

One major issue that needs to be resolved before the government can move on to promote environmental reporting and auditing in a big way is the fear that environmental audit reports may be used against the company involved (Quazi, 1999). That is, the content of an audit may be used as evidence of violations and proof of prior knowledge of violations. An environmental audit may contain self-incriminating evidence of wrongful handling and disposal of hazardous waste, as well as noncompliance with emissions standards. Moreover, there remains the possibility of uncovering environmental problems that are technically unsolvable or financially infeasible to ameliorate with current technology. The liability associated with such negative environmental information may tarnish the reputations of corporations involved (Lang, 1999).

Overcoming this fear of prosecution calls for innovative regulatory enforcement strategies. This may mean giving companies assurances of leniency in situations where, upon self-discovery of violations, companies make good faith efforts to reduce environmental damage and ensure speedy compliance. Good faith effort should be recognized as a strategic defense against strict prosecution and heavy penalties (Lang, 1999).

Environmental disclosure is only a means to an end. It is a necessary first step to promote greater voluntary and proactive environmental management by industries. With systematic and objective tracking of environmental performance, industries may be encouraged to integrate environmental concerns explicitly into their strategic thinking (Quazi, 2001). They may be encouraged to design products and manufacturing processes that are both economically feasible and environmentally sustainable as well.

In conclusion, the general approach of the Singapore government towards achieving greater environmental sustainability is one that emphasizes technological solutions. That progress has been made was evident when the government announced in July 2004 that another landfill was not going to be necessary. Previous pollution control approaches are now being complemented by innovative solutions to enhance resource conservative and pollution prevention at source. Given its well-attested organizational skills, industrial peace, and a corruption-free political and business culture, Singapore may well become a model of an innovator, adopter, and disseminator of environmentally sustainable policies and solutions.

REFERENCES

ACCA (2002a). The State of Corporate Environmental Reporting in Singapore. Certified Accountants Educational Trust, London, www.accaglobal.com/pdfs/environment/tech-ers-001 accessed on June 4, 2004.

ACCA (2002b). The State of Environmental, Social and Sustainability Reporting in Hong Hong. Certified Accountants Educational Trust, London, www.accaglobal.com/pdfs/environment/other/tech-erh-sum.pdf accessed on June 4, 2004.

Bai R. and Sutanto M (2002). The practice and challenges of solid waste management in Singapore. *Waste Management* 22: 557-567.

BCA (2002). Speech by Mr Lim Swee Say, Minister for the Environment at NEA-BCA Seminar-cum-exhibition on environmental management: Best practices for the construction industry at CITI on 26 July 2002, www.bca.gov.sg/newsroom/speech_260702.html accessed on February 4, 2004.

BCA (2003). Speech by Mr Koo Tsai Kee, Senior Parliamentary Secretary for National Development at the opening of BAUCON ASIA 2003 on 19 November, 2003, Singapore Expo Http://www.bca.gov.sg/newsroom/speech_191103.html accessed on June 3, 2004.

BCA (2004). Environmental management. http://www.bca.gov.sg accessed on May 5, 2004.

Canadian High Commission, Singapore (2004a). Update on Singapore's Food Retailing Industry – 2001. A Briefing for Canadian Exporters of Food, Beverages and Agrifood Products. atn-riae.agr.ca/asean/e3243.htm accessed in June 2, 2004.

Canadian High Commission, Singapore (2004b). Update on Singapore's Food Service Industry – 2001 A Briefing for Canadian Exporters of Food and Beverages, http://atn-riae.agr.ca/asean/e3244.htm accessed on June 2, 2004.

EDB (2004). New Initiatives boosts Singapore as testing-bedding hub. Singapore Economic Development Borard, Http://www.sedb.com accessed on July 14, 2004.

ENV (2000). Ministry of the Environment News Release no: 161/2000 22 Nov. 2000. New Features In Singapore's Largest And Newest Incineration Plant Http://app.nea.gov.sg/cms/htdocs/article.asp?pid=580 accessed on January 3, 2004.

ENV (2001). ENV News Release No. 60/2001, April 29, 2001. Speech by Mr Lim Swee Say, Acting Minister for the Environment and Minister of State for Communication and Information Technology - Launch of SEMAC Pte Ltd's new Refuse Collection and Recycling services.

ENV (2002a). ENV News Release No: 025/2002, 02/03/2002. Joint press release by ENV and JTC on JTC-ENV green event 2002, "Towards environmental sustainability: Industries can play a part," 5 March, the JTC Summit, app.env.gov.sg/press.asp?id=SAS331 accessed June 10, 2004.

ENV (2002b). Environmental News Release No: 013/2002, 29/01/2002. Use of incineration bottom ash for road construction – road trail (Information Paper), http://app.env.gov.sg/press.asp?id=SAS318 accessed on June 10, 2004.

ENV (2002c). New features of largest and newest incinerator, http://app.nea.gov.sg/cms/htdocs/article.asp?pid=580 accessed on January 3, 2004.

ENV (2002d). The Singapore Green Plan 2012. Beyond clean and green towards environmental sustainability. Ministry of the Environment.

ENV (2003a). ENV News Release No: 40/2003, November 11, 2003. Speech by Mr. Lim Sweesay, Minister for the Environment at the launch of industrial waste recycling programme for JTC industrial estates at Kallang Basin Industrial Estate.

ENV (2003b). ENV News Release Number: 26/2003, Sept 10, 2003. Speech by Mr. Lim Swee Say, Minister for the environment at the opening ceremony of Sembenviro Alex Fraser Construction and Demolition Materials Recovery Facility at Sarimbun Recycling Park on Sept. 10, 2003.

Feedback Unit. 2002. http://app.feedback.gov.sg/asp/let/let01b.asp?replyId=228 accessed on February 4, 2004.

Fehr, M; Calcado, M.D.R.; and Romão, D.C. (2002). The basis of a policy for minimizing and recycling food waste. *Environmental Science and Policy* 5, 247-253.

Fong, K.P.Y. and Tan, H.M. (2000). Isolation of a microbial consortium from activated sludge for the biological treatment of food waste, *World Journal of Microbiology and Biotechnology,* 16, 441-443.

Kaur, S. (2003). Taking the ash-filled road to total waste recycling. *The Straits Times,* February 26.

Lang, J. C. (1999). Legislative, regulatory, and Juridical dilemmas in environmental auditing. *Eco-Management and Auditing,* 6, 101-114.

Lau, E (2000). Waste incineration to be privatised in five years' time. *The Straits Times,* November 26, Singapore Press Holdings Limited.

Lee, J. and Win, H.N. (2004). Technology transfer between university research centers and industry in Singapore. *Technovation,* 24, 433-442.

Mccoy, P. (2001) Singapore ranked 8th for eco-friendly performance. *The Straits Times,* January 22, Singapore Press Holdings Limited.

Natsteel, (2003). Natsteel will be recycling its steel making b-product, ladle furnace slage, into a value added product, www.natsteel.com.sg/announcements_20022003.htm accessed on July 12, 2004.

NEA. (2002a). The need for waste minimization, app.nea.gov.sg/cms/htdocs/article.asp?pid=1459 accessed on January 3, 2004.

NEA (2002b). How to lower business costs while being environmentally friendly, http://app.nea.gov.sg/cms/htdocs/category_sub.asp?cid=148 accessed on March 3, 2004.

NEA. (2002c). ISO 14000 in Singapore, http://app.nea.gov.sg/cms/htdocs/category_sub.asp?cid=146 accessed on 3 January, 2004.

NEA (2002d). New ideas for better environment management unveiled at the NEA-BCA Seminar. Http://app.nea.gov.sg/cms/htdocs/article.asp?pid=2027 accessed on June 3, 2004.

NEA (2002e). Solid waste management, Http://app.nea.gov.sg/cms/htdocs/category_sub.asp?cid=75 accessed on August 8, 2004.

NEA (2002-2003). *Annual Report 2002-2003.*The National Environment Agency.

NEA (2003). News release no: 07/2003, 25 February 2003. NEA: Successful Trial On Using Incineration Bottom Ash For Road Construction, http://app.nea.gov.sg/cms/htdocs/article.asp?pid=1951 accessed on June 10, 2004.

NEA (2004a). News Release no: 26/2004, 5 Jun 04. ENV Minister Presents The Inaugural SGP 2012 Awards, http://app.nea.gov.sg/cms/htdocs/article.asp?pid=2404 accessed in June 8, 2004.

NEA (2004b). http://app.nea.gov.sg/cms/htdocs/article.asp?pid=982 accessed on February 4, 2004.

NTU (2003). Recycling of steel making by-product into value-added product. Http://www.ntu.edu.sg/cee/staff/environmental/academic/extra/Revised%20Press%2019%20Feb%202003-3.pdf, accessed on May 4, 2004.

Ofori, G. (2000). Greening the construction supply chain in Singapore, *European Journal of Purchasing and Supply Management,* 6, 195-206.

Ofori, G., Briffett, C., Gang, G. and Ranashinghe, M. (2000). Impact of ISO 14000 on construction enterprises in Singapore. *Construction Management and Economics,* 18, 935-947.

Perry, M. and Teng, T. S. (1999). An overview of trends related to environmental reporting in Singapore. *Environmental Management and Health,* 10(5), 310-320.

Quazi, H. (1999). Implementation of an environmental management system: the experience of companies operating in Singapore. *Industrial Management and Data Systems,* 99/7, 302-311.

Quazi, H. (2001). Sustainable development: Integrating environmental issues into strategic planning. *Industrial Management and Data Systems*, 101/2, 64-70.

Show, K.Y., Tay, J.H., and Goh, A. T. C. (2003). Reuse of incinerator fly ash in soft soil stabilization, *Journal of Materials in Civil Engineering*, 15/ 4, July/August, 335-343.

Singapore Environment Council. (2004). Singapore Environmental Achievement Award, http://www.sec.org.sg/seaa/background.htm accessed on January 3, 2004.

Spring Singapore (2004). Local enterprise technical assistance scheme, http://www.spring.gov.sg/portal/products/assist/edf/letas.html accessed on January 3, 2004.

Tan, T.K., Ofori, G. and Briffett, C. (1999). ISO 14000: Its relevance to the construction industry of Singapore and its potential as the next industry milestone. *Construction Management and Economics,* 17, 449-461.

Teh, J. L. (2004). This is where waste food ends up. *The New Paper*, February 3, p. 13.

Wang, J.Y. Olena, S, Tay, S.T.L. Ivanov, V. and Tay, J.H. (2003). Intensive bioconversion of sewage sludge and food waste by bacillus thermoamylovorans. *World Journal of Microbiology and Biotechnology,* 19, 427-432.

INDEX

A

B

C

D

E

F

G

H

I

J

K

L

O

P

Q

R

S

T

U

V

W

Y

Z